Food Service Equipment

THIRD EDITION

D1042508

THIRD
EDITION Food

ANNA KATHERINE JERNIGAN, M.S., R.

Service
Equipment

LYNNE NANNEN ROSS, Ph.D., R.D., L.D.

 IOWA STATE UNIVERSITY PRESS / AMES

158786

ANNA KATHERINE JERNIGAN, M.S., R.D., *is an experienced teacher, administrative and consultant dietitian, and is former director of Nutrition Serivce, Iowa State Department of Health.* She has been consultant author for the American Hospital Association and has served as consultant to the Health Care Finance Administration of Health and Human Services.

LYNNE NANNEN ROSS, Ph.D., R.D., L.D., is president of Creative Concepts, a food service consulting firm. Dr. Ross teaches all phases of food service management to dietitians and managers throughout the United States.

© 1989, 1980, 1974 Iowa State University Press, Ames, Iowa 50010
All rights reserved

Composed by Iowa State University Press
Printed in the United States of America

No part of this book may be reproduced in any form or by any electronic or mechanical means, including information storage and retrieval systems, without written permission from the publisher, except for brief passages quoted in a review.

First edition, 1974
Revised printing, 1976

Second edition, 1980
Second printing, 1983

Third edition, 1989

Library of Congress Cataloging-in-Publication Data
Jernigan, Anna Katherine.
    Food service equipment / Anna Katherine Jernigan, Lynne Nannen
Ross.—3rd ed.
        p. cm.
    Bibliography: p.
    Includes index.
    ISBN 0-8138-0551-1
    1. Food service—Equipment and supplies. I. Ross, Lynne Nannen.
II. Title.
TX912.J47 1989
642'5'028—dc19
                                                                        88-19878
                                                                        CIP

642.57
J55f8

CONTENTS

PREFACE

In order to purchase the kind of equipment that will be most useful in any given facility, it is important to have as much information as possible, for food service equipment has been improved greatly by the computerization of controls, increased use of insulation, and more efficient use of energy. The authors have visited many equipment factories, have studied company specifications for their equipment, and have found that much of the information about equipment is vague. The authors have also worked with architects, engineers, and contractors to develop plans for several hundred health care facilities, schools, restaurants, penal institutions, and even a ship. They have been involved in checking layout efficiency as well as problem areas.

This book is a result of the accumulated information from the above experiences and should be of value to architects, contractors, administrators, dietitians, managers, and others who may be involved in remodeling a facility, replacing equipment, and/or improving the efficiency of food service departments in restaurants, schools, hospitals, and nursing homes. Teachers who are providing information on food service equipment and the basic principles of layout should also find this book of value.

Food Service Equipment

THIRD EDITION

1

Planning Goals: Building, Remodeling, Expansion

The goal in planning and equipping a new or remodeled food service facility is to design, as nearly as possible, an ideal system within the basic limitations of available funds, current technology, and the environment.

The first step is to prepare a work-flow plan. The facility must have easy access to the outside for receiving supplies and to the main part of the building for distributing food and beverage. Food and supplies should move from receiving and storage to preparation and service then to warewashing in a smooth, orderly manner without crossing and backtracking. The layout of work areas, allowing space for all equipment to be added, should create a work flow with minimum traffic in each section but with maximum accessibility to the service areas. The internal food service layout should be designed for optimum production at minimum cost, taking into account the food products and procedures to be used.

The relationship of one area to another and the logical sequence of work in each individual area relate to the total flow of work through the facility. The major objectives of a good plant layout are

To avoid unnecessary capital investment
To provide effective space utilization
To integrate all factors affecting the layout
To accomplish a work flow through the plant

To develop a flexible arrangement that can be readily read-
justed
To simplify the production process
To stimulate effective labor utilization
To ensure the safety and satisfaction of workers[1]

Once the work flow has been established, the space al-
lotment and equipment required for each area must be
planned. Choose equipment and utilities carefully to ensure
that they are best suited to the type and scope of service
desired and that their design and installation permits ready
accessibility for proper maintenance.

Compactness does not necessarily make for an efficient
layout. Many food service departments hold the maximum
amount of equipment in the minimum amount of space, but
this compactness may not save steps for the cooks and may
even cut down on their productivity. Cooks need ample work
space adjacent to equipment for placing products immediately prior to and after cooking.

Equipment

There are many variations of the four basic types of menus,
each of which determines the kind and amount of equipment
needed in a facility (Table 1.1).

In a facility where service is continuous over a span of
several hours, many of the hot foods will need to be cooked
as quickly and as near serving time as possible. For maximum quality and no waste, a jet steamer, a convection oven,
and/or small steam-jacketed kettles are appropriate. At the
other extreme in a food service where large quantities of
food are needed in a short period of time, such as a school, a
compartment steamer, large ovens, and large steam-jacketed
kettles are necessary.

EQUIPMENT FOR CONVENIENCE FOODS SYSTEMS. Many
convenience foods are already in use; therefore, meat saws,

---

[1]Geraldine M. Montag. "Development of Overall and Detailed Layout."
Selected Papers from Food Administration Conference, Work Design—Equipment
and Layout, July 20–23, 1966, pp. 12–20.

TABLE 1.1. **Kind and amount of equipment needed for selected facilities**

Menu	Production Schedule	Equipment of Choice
STEAK HOUSE/Waitress Service		
Soup	Prepared ahead	Steam-jacketed kettle
Steaks, chops	Cook to order	Grill and broiler
Salad and dressings	Prepared ahead	Refrigerator
Baked potato	Staggered baking	Convection or microwave oven
Rolls	Staggered baking	Convection oven
Butter	Ready to serve	Refrigerator
Ice cream	Ready to serve	Freezer
HEALTH CARE FACILITY OR RESTAURANT/Cafeteria Service		
Soup	Prepared ahead	Steam-jacketed kettle
Roast meat	Finish ½ hr before serving	Convection oven
Casserole	Staggered baking	Convection oven
Two vegetables	Cook to order	High pressure or convection steamer or oven
Two salads and dressings	Prepared ahead	Refrigerator
Bread	Ready to serve	Room temperature storage
Butter	Ready to serve	Refrigerator
Two desserts	Prepared ahead	Steam-jacketed kettle or oven or freezer
SMALL HEALTH CARE FACILITY OR RESTAURANT/Limited Menu		
Entree	Finish at serving time	Convection oven or range
Vegetable	Cook to order	High pressure or convection steamer
Fruit or salad	Prepared ahead	Refrigerator
Bread	Ready to serve	Room temperature storage
Butter	Ready to serve	Refrigerator
Dessert	Prepared ahead	Oven or refrigerator or freezer
SCHOOL FOOD SERVICE OR RESTAURANT/Fast Food		
Entree	Finish at serving time	Oven or steam-jacketed kettle
Vegetable	Staggered production	Compartment steamer
Salad	Prepared ahead	Refrigerator
Bread	Ready to serve	Room temperature storage
Butter	Ready to serve	Refrigerator
Dessert	Preapred ahead	Oven or steam-jacketed kettle or refrigerator or freezer

butcher blocks, potato peelers, large vegetable preparation areas, and baking areas are being reduced or omitted from food service departments.

If a total convenience system or prepared frozen entrees and baked goods are used, the preparation equipment necessary would include worktables, counters, and sinks for salad preparation; steamers, convection ovens, and microwave ovens for reconstituting frozen food; refrigerators for holding prepared food, salad ingredients, milk, and leftovers;

freezers for frozen food; and carts, serving equipment, dishes, utensils, and dishwashing equipment for serving and cleanup.

EQUIPMENT FOR CATERED FOOD SYSTEMS. If catered hot food is delivered to a satellite unit in heated containers and cold food is delivered in cold containers, a serving line and dishwashing equipment would be the main items needed for this system. Other equipment may be a refrigerator for butter or spread, eggs, fruit juice, milk, and items that must be kept cold; a freezer for ice cream; coffee makers and a source of water for tea; a blender to make special nourishments for emergency use; equipment for poaching, frying, and preparing soft-cooked eggs, as well as for making toast; space for storing bread and other extra food items; an ice machine; and perhaps a microwave oven.

If contract food service is to be used, it is important to have an agreement in advance to determine who will be responsible for the replacement, maintenance, and repair of equipment.

Work Centers

In developing work-flow patterns, each work center must be identified and located in relation to the other centers.

DOCK. The loading dock should be as high as the floor of the average delivery truck. In some cases this will necessitate excavating a drive so that the truck will be below dock level. An acceptable alternative is a motorized lift that is approximately 3 × 6 ft, which can be located outside the building and to the side of the service entrance. The motor is located below the dock and covered when not in use. A roof high enough to allow service trucks to drive under should extend over the dock area.

RECEIVING AND STORAGE AREAS. Receiving must be adjacent to the loading dock or service entrance. Plan the main storage near the receiving area and in a logical line of routing to the kitchen storage. The ideal location for daily storage is adjacent to the preparation area on the same floor level. If storage is on another level, a lift or elevator is necessary.

LOW-TEMPERATURE STORAGE. Storage refrigerators and freezers should be near where food is received or enters the production area. They are needed in all units and should be accessible to each work area.

COOK'S UNIT. The cook's unit is the main production area. Work-flow lines to and from this area should be as short as possible since raw materials come to this area from the meat and vegetable preparation sections, dry and low-temperature storage areas, and by direct delivery. It is usually desirable to have the cook's unit close to the pot-and-pan washing area as well as in the center of the room with the other preparation areas located around it.

BAKER'S UNIT. The baker's unit may be somewhat more remotely located because of the ease of transporting raw materials and finished products. In small units it is generally desirable to locate this unit adjacent to the cook's unit so ovens, sinks, mixers, and other equipment may be shared.

SALAD UNIT. A salad unit is best situated so that the routing is in a straight line from the supply of raw materials to the serving area. This provides access to the chilled salads at the point of service with a minimum of transportation.

CAFETERIA OR SERVING AREA. The cafeteria or serving area must be adjacent to supporting areas to permit a continuous supply of food. Locate dishwashing areas so that dirty dishes from the dining room can be returned by a direct line or by means of a conveyor belt that is readily accessible to the guests.

WAREWASHING AREA. The warewashing area should be easily accessible from the cook's and baker's units and remotely tied in with the cafeteria or serving area.

Carefully choose all equipment to ensure that it is best suited to the type and scope of service desired and that installation permits ready accessibility, including utilities, for proper maintenance. Be sure the drainboards are adequate and enough racks are available so that all pots and pans are correctly stored.

Criteria for Evaluating Preliminary Floor Plans

SPACE.

1. Adequate space for needed equipment should be provided; space that serves no useful purpose should be eliminated.
2. Adequate aisle space (3½ ft in work centers, 4 ft where oven or refrigerator doors open, 6 ft for normal traffic aisles) should be provided.
3. Space for complete work centers appropriate for the type of facility being designed should be provided so that employees have everything available with a minimum of walking.
4. Space for additional storage, refrigeration, a small amount of cooking equipment, and carts to provide for expansion is desirable. If this is done properly, plumbing and wiring should not have to be changed. A good arrangement for warewashing that will allow for increased volume must be part of the original design.

WORK-FLOW PATTERN.

1. Four basic doors and access to two corridors are necessary for efficiency.
   a. An entry through a corridor for delivery of produce and supplies, either from the outside or from other storage areas.
   b. A corridor or an elevator for moving completed trays of food for the patients/residents from the serving area.
   c. A serving area (cafeteria) to provide food to the staff.
   d. A door and/or window to return soiled trays and tableware from patient/resident and staff dining areas.
2. Continuous flow from receiving to prepreparation, or preparation (depending on the level of convenience to be

used), to serving to warewashing without crossing paths is necessary.
3. Good separation of clean and dirty areas is necessary for sanitation as well as efficiency of operation.

Expansion and/or Remodeling

Expansion of the food service area should increase the original area, be indicated by dotted lines on the original blueprint, and be written in narrative form for future reference. Plumbing and electrical raceways should be designed to accommodate any expansion program. The preparation area should be adequate for present needs yet allow for the addition of one or two extra pieces of cooking equipment, such as a steam-jacketed kettle, tilting braising pan, or compartment steamer. Counter space, stationary tables with sinks attached, as well as dishwashing and warewashing areas should be adequate for expansion. Mobile work stations, mobile dish-storage equipment, carts, and refrigeration can be added if the original plan was carefully developed with possible expansion in mind.

Remodeling maintains the original space but should eliminate old equipment that will be replaced by newer more versatile equipment; relocate usable equipment that is improperly placed, even if this requires the addition of new utility lines; and relocate less-usable items. Priority should be given to developing a good work-flow pattern and complete work stations. It is very important to provide electric raceways with access to current, adequate lighting, and adequate plumbing and sewer lines. Flexibility is the key to making any program work.

The following plans (A, B, C) are examples of expansion possibilities. They are the result of efforts to make the original plans more efficient and handle greater needs for refrigeration, freezer, and storage space.

EQUIPMENT LIST FOR PLANS A, B, C

1. Prerinse and disposer
2. Disposer
3. Ceiling-mounted hose reel
4. Elevated shelf
5. Soiled-dish table
6. Quick drain
7. Single-tank dishwasher
8. Clean-dish table
9. Cold food unit
10. Tray carts
11. Utility rack
12. 2-compartment pot washing
13. Coffee urn
14. Counter
15. Drawer
16. Milk dispenser
17. Bread storage
18. Toaster
19. Worktable
20. Hot food unit
21. Dish dispenser
22. Service sink, janitor's closet
23. Receiving table
24. Reach-in refrigerator, 50 ft³
25. Reach-in refrigerator, 75 ft³
26. Pass-through refrigerator, 50 ft³
27. Reach-in freezer, 50 ft³

28. Reach-in freezer, 75 ft³
29. Compartment steamer
30. Hood
31. Range, 8 open burners, griddle, and 2 ovens
32. Range, 4 open burners, griddle, and convection oven
33. Slicer
34. 20-qt mixer
35. Sink(s) with or without disposer
36. 2-compartment sink
37. 3-compartment pot-and-pan sink and disposer
38. Handwash sink
39. Shelving
40. Desk
41. Ice cream cabinet
42. Cup dispenser
43. Ingredient bins
44. Pass window
45. Utility cart
46. Ice machine
47. Convection oven
48. Tilting braising pan
49. Steam-jacketed kettle
50. Tray assembly

PLAN A. The space available (32 × 25 ft = 800 ft²) serves an 80- to 100-bed facility (Fig. 1.1).

Assessment

1. There is only one service corridor; therefore carts to patients' rooms have to pass through the dining area.
2. The pot-and-pan sink has only 2 compartments. Most state health departments require 3 compartments.
3. There is no pot-and-pan rack.
4. The work space for cooks is inadequate, for that area doubles as a hot food serving table.
5. The mixer is not convenient to the cook's area.
6. The janitor's closet is located so that waste material has to pass through the clean work area.

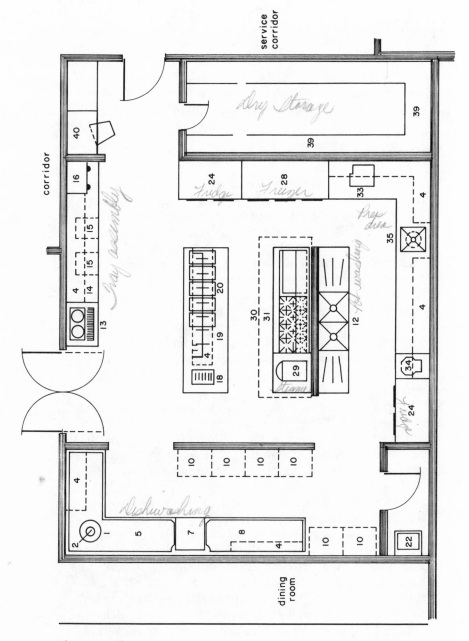

Fig. 1.1. Plan A.

7. There is no separation of clean- and soiled-dish areas in the dish room.
8. There is no handwash basin.
9. The refrigerator adjacent to the janitor's closet is inconvenient for use in setting up trays.

PLAN B. The space available (40 × 42 ft = 1,680 ft²) serves a 100-bed hospital or 150-bed skilled-care facility with 50 persons coming to the dining area for meals (Fig. 1.2).

Assessment

1. The work-flow pattern is good for it allows continuous movement forward with very little back-tracking of steps.
2. There is a little shortage of storage space, especially if quantity purchasing is desirable. Additional storage space for stacks of cases might be provided elsewhere.
3. The cooking island, work space, and pot-and-pan rack are quite adequate. An effort should be made to see that the cook's and baker's work centers are complete with drawers, bins, cabinets for condiments and spices, and racks for utensil storage.
4. The tray line or serving area is a little small (18 ft in width is barely adequate when a refrigerator is included in this section). The space would offer more possibilities with a width of 20–22 ft.
5. The tray cart storage space is too small if it should be necessary to provide 5 or 6 carts instead of 4.
6. The refrigeration space is not adequate for more than 100 beds and not enough to support that many beds and serve an equal number of employees. Additional refrigeration is needed if many fresh fruits, vegetables, eggs, and meat are to be kept on hand. If most of the items listed are frozen, more freezer space is needed. Space for the milk needed should be calculated, and either milk dispensers or storage space for milk should be provided.
7. The dishwashing area should be adequate. A sink for handwashing is easily available. The clean-dish area is spread in the form of an obtuse angle, so that there will be space for mobile dish-storage carts and tray-delivery tables. The person handling clean dishes has accessibility to the tray line and therefore could clean the serving equipment if there is more time than is needed for putting away racks of clean dishes.
8. The office space could be more elaborate; however, the desk is accessible from the corridor without moving through the clean work areas.

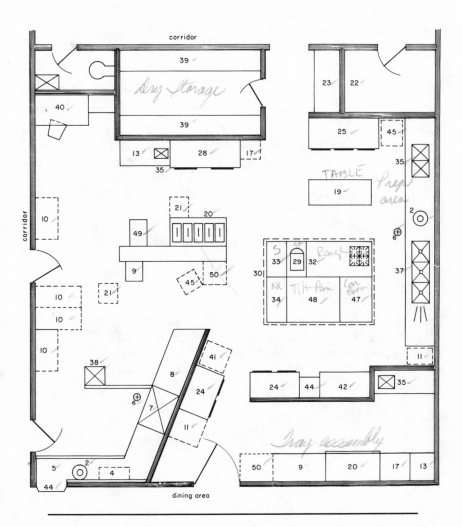

Fig. 1.2. Plan B.

PLAN C. The space available (25 × 28 ft = 700 ft²) serves a 70- to 90-bed facility (Fig. 1.3).

Assessment

1. Storage space, the janitor's closet, the freezer, and the refrigerator are all near the service corridor for deliveries.
2. One disposer is accessible for the vegetable sink and the pot-and-pan sink; however, drain space for the pot-and-pan sink is short and no pot-and-pan rack is provided. A 20-in.-wide pot-and-pan rack could be located behind the range.
3. A future refrigerator is shown near the cook's table. This unit would be needed when the facility reaches capacity. A 10-gal self-contained steam-jacketed or high-pressure steamer could be added by the range if the mobile clean-dish storage unit could be stored under the clean-dish counter.
4. There is a separation of dirty and clean areas, especially between that for dirty dishes and the serving area. There is also a good flow of dirty dishes into the dishwashing unit and then clean dishes to carts to be moved back to the area of use.
5. All items necessary to serve trays or the dining area are located adjacent to that area.

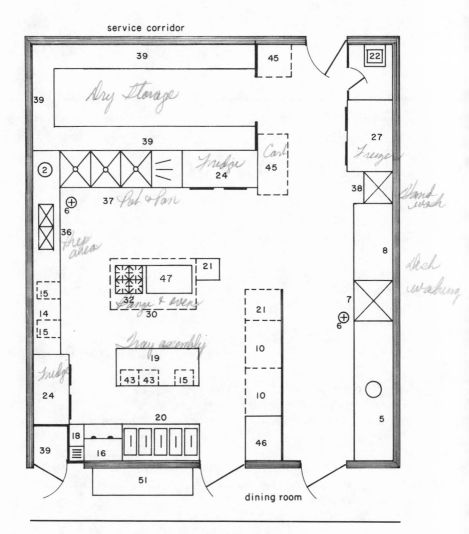

Fig. 1.3. Plan C.

Construction Factors

2

Before any equipment can be operational, the gas, electricity, plumbing, ventilation, and fire control must be considered and planned carefully.

Gas — Natural and Propane

Many food service managers think it is wise to have part of the cooking and serving equipment energized by gas. Gas supplies a source of immediate heat and when turned off, the heat stops immediately. If at least some of the equipment is gas powered, it is still functional in an electrical failure and the menu can be modified more easily to adapt to the emergency. However, many gas-fired burners are triggered by an electrical solenoid thermostat so it is important for some auxiliary power to be available in case of a power outage.

Food service equipment designed to use gas may be less expensive than the same equipment that uses electricity. The cost of operating gas-fired equipment should be compared to the cost of electricity for an equal amount of heat or energy. Planners and designers must decide which pieces of equipment will use gas and which will use electricity.

Examples of equipment that might be powered by gas are

Tilting braising pan	Dishwashers
Oven	Ranges

Broiler   Griddles
Deep-fat fryer

Because natural gas is piped directly from the supplier to the consumer and is not stored on the premises, there is the possibility that the supply will not meet the demand during periods of peak usage; thus, it is good to have a propane-air standby unit installed. The dealer can attach the necessary regulators so that the mix of fuel and air will perform with the same orifice setting as with natural gas. If natural gas is not available, burners can be adjusted to use straight propane, which has a higher British thermal unit (Btu) rating.

It is important for the mechanical engineer or contractor to calculate the necessary power output of all utilities. This calculation or estimate should be based on maximum load so that during peak periods there will be an adequate fuel supply for the equipment to perform to capacity. The total Btu load must be determined to properly size the storage container, regulator, and piping. Whoever provides the basic input into the planning might project the total Btu's needed by totaling the requirements of each piece of equipment, allowing for a 10–15% margin of safety.

A manometer, which indicates the presence of leakage, is used to check the pressure, which should be held to 10 pounds per square inch (psi). There are specific requirements for the size and kind of pipe depending on the distance transported, the flow rate, and the usage volume. Generally, copper pipe is used for shorter distances and lower flow rates and usage volumes, but iron pipe becomes necessary as these factors increase.

Propane storage containers and piping will be installed by the local supplier in accordance with National Fire Protection Association standards.

Electricity

The food service manager should provide information as to the location and number of electrical outlets needed, and the level of current necessary. The availability of power in the correct amount and kind is important to assure the correct

functioning of the equipment. An electrical engineer can provide valuable assistance but his services may not be necessary.

Examples of equipment in the preparation area that
might be electrical are

Oven	Freezer
Griddle	Refrigerator
Broiler	Mixer
Deep-fat fryer	Blender
Surface cooking units	Slicer
Electric knife	Can opener
Food waste disposer	Shredder or chopper
Mobile floor-cleaning unit	Tilting fry kettle

Examples of equipment in the serving area that might be
electrical are

Ice machine	Clock
Ice cream cabinet	Toaster
Tray delivery carts	Coffee maker
Hot food serving table	Tray conveyor
Heated dish dispensers	Refrigerator
Oven for heating disks or pellets	Milk dispenser
Unit for heating water (to make hot beverages)	Feezer

Examples of equipment in the cleanup area that might
be electrical are

Mobile floor-cleaning unit	Dishwasher controls
Soiled-dish conveyor belt	Disposer

A power pole or a conduit located in a raceway that is
attached to the ceiling is a good way to deliver electrical
current for the equipment needed on the serving line. To
provide current to the tray conveyor unit, attach a removable
drop cord with locking connections. Receptacles or outlets
for 120-volt single-phase and 208-volt single- and three-phase
current may be provided through factory-wired circuits in
the tray conveyor unit. It is necessary to determine in advance which side of the conveyor unit will need the various
levels of current. Calculate the total power needs of the tray
conveyor unit by totaling the individual load requirements of
the various components. Additional circuits and capacity
should be available, also.

Specify the voltage desired for each piece of equipment. If a 220-volt piece of equipment is plugged into a 208-volt line, it will not operate at the maximum rated capacity (it will not get as hot as it should). If a 208-volt piece of equipment is plugged into a 220-volt line, it may burn out.

Three-phase current is far more efficient and requires a much smaller wire for the same current than for single-phase current. Use a five-pin receptacle configuration on three-phase outlets (Fig. 2.1). This configuration consists of three-phase wires (X,Y,Z), a neutral wire (N), and an equipment ground (EQGR).

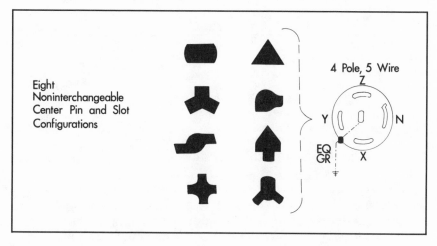

Fig. 2.1. Five-pin receptacle configurations.

Specifications for electrical equipment should include the wattage and/or horsepower of the unit as well as voltage characteristics, phases, and plug types. A mechanical engineer should be responsible for this if one is on the job, but whoever specifies the equipment should see that the job is done properly.

There are many special pin configurations. Figure 2.2 shows examples of receptacles with three-pin configurations. Receptacles can be installed to match the plugs on the appliances.

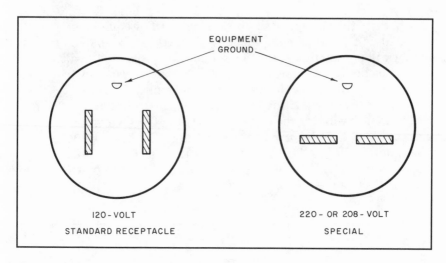

Fig. 2.2. Three-pin receptacle configurations.

When food carts need special receptacles, the receptacles should be specified by the manufacturers' number (e.g., the Hubbell or Arrow Hart numbers). The locations for the carts also should be designated so that the correct receptacles can be provided. In some cases it may be desirable to change the plugs so that several pieces of equipment may plug into particular receptacles interchangeably. The pin configuration of all receptacles and plugs supplying like voltages should be the same for maximum flexibility and minimum confusion.

Protect all electric motors by individual overcurrent protection devices, such as circuit breakers. Also protect motors against overheating as a result of clogged air ducts, low voltage, or high room temperatures. A burned out motor is not only costly to repair, but the loss of use of the appliance is inconvenient.

When selecting electrical equipment for a food service facility, follow three basic guidelines:

1. Find out what system voltage is available.
2. Purchase equipment that will match the system voltage.
3. Adapt existing equipment that was designed to operate on a different voltage by use of an auto-booster transformer.

Plumbing

Floor drains should be located appropriately in the dish-washing area; near the pot-and-pan area; and near the steam-jacketed kettles, steamers, and vertical cutter/mixers. The floors should slope toward the drains. When a heavy load of liquid is to be disposed of and the floor needs to be fairly level, a drain can be dropped below the floor level and covered with a removable grill. This design is very helpful in areas where there is a great deal of cart traffic.

A supply of both hot and cold water must be available at the dishwasher, the pot-and-pan sinks, the coffee urn, sinks in the preparation area, and handwashing basins.

Facilities that plan to install vegetable peelers, vertical cutter/mixers, steam-jacketed kettles, and other similar equipment should have water and drains located where the equipment is installed.

Ventilation

A good ventilation system, capable of removing all the waste products of combustion and cooking by allowing proper operation of the exhaust system, will increase the efficiency of the employees and life expectancy of the equipment. Since ventilation in food service areas cannot be independent of ventilation in other parts of the building, it is important to employ a good mechanical engineer to develop this plan. With knowledge of where hot food will be held and served and where clean dishes will be drying, the engineer can de-sign the system to eliminate any cold air being pumped into these areas. It is important to bring clean heated air in over the clean-dish table and to exhaust the air above the soiled-dish table.

The area in which clean tableware is handled and stored is a particular concern in control of contamination. A U.S. Public Health Service study of dishwashing areas in 21 health care facilities in the Minneapolis–St. Paul area in 1968 indicated that the type of air-handling system was the only factor that had an effect on the amount of microbial contamination in the air.[1] Contamination was lower in facili-

[1] Walter H. Jopke and Duane R. Hass. "Contamination of Dishwashing Facilities." *Hospitals,* 1970 (March 16):124–217.

ties that had air-cleaning devices in the ventilation system (i.e., facilities that were air-conditioned). Hoods designed to provide good ventilation also increase the general efficiency of the dishwashing operation by reducing humidity and heat.

Ventilation is no longer just an exhaust fan, and well-designed ventilators can reduce energy cost. Fans are being designed that exhaust a special amount of air as required by building and safety codes. There is also a method available that mixes exhausted kitchen air with outside air, cleans it, and then reintroduces it into other areas of the facility as warm makeup air. The makeup air is then drawn by negative pressure into the food service department, back through the hood and up the exhaust, and then the process starts over. The key to this system is a 2-section range hood with both an exhaust and a makeup air compartment. Instead of sucking high amounts of expensively heated or cooled air from inside the facility, the system introduces makeup air directly from outside.

There should be dampers in the duct system of hoods to control the rate of air exhaust. The rate of airflow is directly associated with the efficiency of the system: the longer the pipe or the more turns it has, the larger it must be or the greater the pressure required to maintain the optimum air-flow rate. Air must be supplied around the peripheral areas of the hoods to avoid drafts in other areas of the kitchen. If the installation of hoods in an existing building is not possible, there is a ventilation system available that can be vented through an outside wall or floor.

A self-cleaning system in the hood over the cooking area is a labor-saving device. However, if a self-cleaning system is too expensive, consider designing the inside portions of the hood so that air filters can be removed easily and washed in the dishwasher or by steam under pressure.

Production areas should be maintained at a temperature of 75–80°F (24–27°C) and the humidity level of 50–60% for maximum efficiency of the employees.

Storage areas should also be well ventilated. A storeroom without windows may be cooler than one with windows, which permit lots of sunlight. The storeroom that is not air-conditioned should at least have louvers at the top and bottom of the doors for ventilation.

There is a common statement that says that the rate of a chemical reaction doubles for each 18°F (10°C) rise in temperature. Applying this rule to products such as canned foods that lose quality only by slow chemical reaction means the general quality loss at 78°F (26°C) storage would be about twice the quality loss for the same product stored at 60°F (15°C) for the same period of time. To reduce quality loss of canned products to a minimum, the temperature of the storeroom should be maintained between 50°F and 70°F (10°C and 21°C). Motors for remote refrigeration should never be in the storage areas because heat from the motors can raise temperatures and cause deterioration of canned goods. In addition there should be no overhead waste or water pipes that may leak and contaminate the foods.

Floors, Walls, Ceilings, and Lighting

Efficiency is improved when floors, walls, and ceilings are selected to optimize lighting and acoustics. Quarry tile is one of the most preferred types of flooring, but a disadvantage is its hardness. However, this can be compensated for by having employees wear shoes with neoprene soles. Floors made of poured concrete and epoxy that are done well and have up to ten layers of hard finish on them are gaining in popularity. (Each layer needs to dry 4 hr before the next layer is applied. If there is more than a 4-hr drying span the material will get too hard and split between layers.)

White or light-colored walls reflect light best, but the walls should not be so smooth that reflected light causes glare. This is tiring and decreases efficiency among employees.

Ceilings should be white or off-white acoustical tile to reflect light and absorb sound. Some state regulations require that ceilings be smooth and cleanable. Light should be 35–50 footcandles on the work surface. Sound should not exceed 63 decibels. To reduce utility costs and provide adequate light, each work area should be properly lighted for the job to be done and excess light in other areas avoided.

Fire Control Systems

Kitchens should be protected with extensions of the fire alarm and/or sprinkler systems, including an automatic fire extinguishing system in exhaust hood assemblies over cooking equipment. These systems should be capable of being activated manually.

Three major types of fires could occur in the food service area. Type A fires are paper, wood, or other combustible materials that can be extinguished with water, CO_2, dry chemical, or foam. Type B fires are grease or other flammable liquids and can be extinguished by CO_2, dry chemical, or foam. Type C fires are electrical and can be extinguished with CO_2 or dry chemical. Every food service should have at least one portable CO_2 or dry chemical fire extinguisher located in the cooking area, which would be effective for all three types of fires.

FIRE SAFETY. The following items should be considered during initial construction as well as remodeling:

1. Regular and emergency exits adequate and unobstructed.
2. Exit lights adequate and bright.
3. Fire doors (solid, self-closing, metal) working properly.
4. Alarm boxes, extinguishers, and other fire-fighting components in proper place, unobstructed.
5. Early-warning detectors and overhead sprinklers working properly. (Allow at least an 18-in. clearance between sprinkler heads and possible obstructions.)
6. Worn cords and plugs, broken switchplates, and other elements repaired or replaced; no long lamp cords, defective electrical equipment, or overloaded electrical lines.
7. All ducts for heating, air conditioning and ventilation, and especially exhaust ducts over kitchen ranges, etc., kept clean.
8. Dampers, which close automatically when their sensors detect fire, in working order.
9. Trash collection areas adequate.
10. Fire-retardant coatings on all materials, even plastic plants, renewed regularly.

11. Noncombustible products, such as metal wastebaskets, nonflammable blankets and bedspreads, and the newer fire-retardant mattress covers.
12. Fire extinguishers accessible but not so close to hazardous areas that it would be dangerous to approach them.

Occupational Safety and Health Act

The Occupational Safety and Health Act (OSHA) provides appropriations for carrying out the act and defines penalties for violations. The act specifies that it be the policy of the state to assure, so far as possible, every worker safe and healthful working conditions and to preserve human resources.

Labels and other appropriate forms of warning necessary to ensure that employees are aware of all hazards to which they are exposed must be provided. Suitable protective equipment and control or technological procedures to be used in connection with such hazards are prescribed.

SAFETY FACTORS. The following list is from the *Safety Manual for Nursing Homes & Homes for the Aged* prepared by the National Safety Council in cooperation with the American Nursing Home Association. These features must be provided at the time of construction to aid in operating a safe facility:

1. Aisles, corridors, exits, and stairs designed so that they can be kept clear at all times.
2. Electrical equipment grounded either through a built-in three-wire system or by attachment of a separate ground wire.
3. Switches recessed or otherwise guarded so that equipment cannot be started by accidentally leaning on or brushing against the switch.
4. Switches located so that employees do not have to lean on or against metal equipment to reach them.
5. Hooks so that mops can be safely stored and aired to dry.
6. Adequate storage space to prevent "temporary" storage in the receiving area or hallways, and aisles wide enough

to give employees room to work.

7. Storage space for equipment, utensils, and supplies needed in work areas.
8. Food service department designed so that refrigerator doors do not swing into and block a major trafficway or corridor.
9. Devices that permit emergency exits, such as a bypass device on the door or an alarm bell in each walk-in refrigerator.
10. Screens or mesh guards around blower fans in walk-in refrigerators.
11. Ventilating systems designed so that grease filters can be easily removed from range exhaust hoods to be cleaned.
12. Adequate electrical circuits to avoid too many appliances from being connected into one circuit.
13. Knife racks in areas of use so that knives can be stored safely.
14. Traffic flow planned so that workers entering the dining room do not collide with others returning from the dining room.
15. Swinging doors marked IN or OUT on both sides to show use.
16. Draperies, blinds, and curtains made of fire-resistant materials, with pictures and drapes securely fastened to walls.
17. Tables placed to allow aisle space for removal of dishes without lifting trays of food over the heads of diners.

Other safety factors to consider:

1. All electrical equipment safely located, maintained, and operated.
2. Each employee instructed in correct methods of lifting and handling various types of containers.
3. Adequate equipment for handling bulk supplies.
   a. For opening crates, barrels, cartons (hammer, wire cutter, pliers, cardboard carton openers).
   b. For moving bulk supplies (hand trucks, dollies).
4. Dollies or other wheeled units for garbage containers to eliminate lifting by employees.
5. Carts designed to hold cleaning supplies.
6. Storage of foodstuffs completely separate from storage

of cleaning powders, insecticides, and other poisonous substances.

7. A safe ladder for reaching high storage areas.
8. A thermometer outside and/or inside each refrigerator to provide easy checking for proper temperature.

Energy Management and Conservation

Any industry should have management that is committed to conserving energy, selecting energy-efficient equipment, repairing equipment when necessary, inspiring employee support, and training employees to use energy wisely.

Food service is an energy-intensive industry. A National Bureau of Standards study estimates heating-energy waste at 40%, cooling-energy waste at 40%, and lighting-energy waste at 15%. The National Restaurant Association, in conjunction with the Midwest Research Institute, completed several studies on the use of energy in the food service industry and produced an in-depth analysis of energy conservation methods and guidelines. Copies of this report, "Energy Management and Energy Conservation Practices in the Restaurant Industry," can be obtained from the National Restaurant Association, 150 North Michigan Ave., Suite 2000, One IBM Plaza, Chicago, Illinois 60601.

EQUIPMENT SELECTION. Energy can be conserved by careful selection of equipment.

1. Air-tight refrigerators that maintain proper temperatures should be selected; specify automatic heat-level cutoff.
2. Low-temperature storage equipment with automatic defrost cycles to epitomize efficiency.
3. Steam-jacketed kettles and/or deck or high-pressure steamers that are self-contained or connected to clean steam to provide efficient use of energy.
4. Timers for automatic control of cooking time to determine doneness and reduce overcooking, which requires more energy.
5. Microwave ovens are very energy efficient; little energy is lost to the oven temperature.

6. Tilting fry pans save power because they are multipurpose equipment.
7. Convection ovens preheat more quickly than conventional and deck ovens.
8. Multipurpose equipment has flexibility and can be used on a continuous basis if needed to do several types of cooking jobs; oversized equipment wastes energy.
9. Hot water heaters should be selected on the basis of the type of dishwasher to be used.
10. Fluorescent lighting is more efficient than incandescent; it gives off four times the brightness for the same voltage and emits less heat, which puts less demand on air-conditioning systems. Energy needed for lighting a food service department can be reduced and lighting needs improved if good lighting is provided over work areas such as work centers, tray lines, dishwashers (especially clean-dish areas) rather than lighting the whole service area from ceiling fixtures that have decreased illumination at work heights.
11. A makeup air unit in the food service department cools air more cheaply in the summer and heats air more efficiently in the winter than regular heating, ventilating, and air-conditioning systems.

Energy usage can be reduced by proper location, use, and care of equipment.

1. Heating equipment should be clustered away from cooling equipment.
2. Hot water lines to booster heaters should be kept as short as possible and should be insulated to keep heat loss to a minimum.
3. Water treatment units should be installed on hot water and steam lines.
4. Eliminate preheating of ovens or preheat with thermostat at desired temperature; they do not preheat faster if set higher.
5. Determine a schedule of preheating times for hot food serving tables, grills, broilers, and fryers. Generally 10–30 min (depending on the appliance) is adequate. Cooks who turn everything on when they start duty, so that it is ready when needed, greatly increase energy costs.

6. Avoid preparing items during hours of peak load current consumption.
7. Check to see that oven thermostats are properly calibrated to assure optimum cooking temperature and maximum efficiency.
8. Load and unload ovens quickly to avoid unnecessary heat loss. Each second an oven is open, it loses about 1% of its heat, or about a 10°F (5°C) heat loss. Energy is needed to raise the temperature again.
9. Turn off cooking and heating units when not needed.
10. Cook meat at low temperatures. Cooking a roast for 5 hr at 250°F (125°C) will reduce the energy consumption by 25–50% compared to cooking the same roast 3 hr at 350°F (175°C).
11. Use a meat probe attached to a gauge outside the oven to reduce heat loss that results from opening the oven door to check doneness.
12. Keep coils on the refrigerator clean. Replace worn out gaskets, cords, and motor parts on refrigerators and other equipment.
13. Use hot tap water for cooking whenever possible; a water heater requires less energy than most cooking tops to heat the same amount of water.
14. Cover pots used on the range top to retain heat.
15. Reduce heat as soon as the product comes to a boil; allow it to simmer according to the recipe.
16. Adjust thermostats on tank heaters to maintain proper temperature control.
17. Check the drain and fill the valves on water lines to make sure they are sealed properly. Leaking valves represent constant hot water consumption, which means wasted energy and inadequate hot water. Maintain a regular schedule for deliming, as chemical deposits reduce heating efficiency.
18. Run only full racks through dishwashers. Scrape and rack dishes during low-volume periods, turn on machine only when several racks are waiting and during peak-load periods, and turn machine off when not in use.
19. Close off dining areas not in use; turn off heat and cooling systems in these areas.
20. Draw draperies in cold weather, when possible, to conserve heat.

Management can reduce energy use through awareness and effort.

1. Establish a committee to formulate a plan for conservation of energy, including goals, guidelines, and execution.
2. Conduct walk-through surveys of the facility, listing each piece of equipment that requires energy input.
3. Determine typical operating hours or actual time equipment is using energy each day. Reduce hours of use when possible.
4. Examine doors, windows, and walls for cracks and worn insulation. Record the condition of each and make recommendations for action to eliminate problems.
5. Check heating and air-conditioning systems for leaks. Eliminate problems as soon as possible.
6. Be sure that hot water and steam lines, and hot water storage tanks, are properly insulated and/or make recommendations for action. (Faucets with spring return handles reduce hot water waste.)
7. Stop hot water leaks. Just 90 drops a minute from a single outlet can result in a loss of 300 gal of hot water per month, plus the energy paid to heat it.

Underwriters Laboratories

Underwriters Laboratories (UL) is a product-testing organization.[2] If a product bears the UL label or seal, it has met and must continue to meet rigid safety standards. These requirements include every part of the appliance that could, through some malfunction, create an electrical shock, hazard, or fire. Participation in the UL program is strictly voluntary, but public awareness of safety has made compliance with safety regulations vital for manufacturers.

National Sanitation Foundation

The National Sanitation Foundation (NSF) is a voluntary organization, which originated through the cooperative ef-

---

2. Underwriters Laboratories, 207 E. Ohio, Chicago, IL 60611.

fort of industry representatives and professional public health officials to develop sanitation standards.[3] It promotes education and conducts research in areas where quality affects the public health.

The seal contains the initials NSF and is granted only after an investigation of the applicant's manufacturing methods. In some cases the investigation may include the testing of the equipment to show compliance with standards. Equipment bearing the NSF seal is evaluated annually or at the request of a purchaser.

Some standards of sanitation must be met:

1. Materials used must be nontoxic.
2. Equipment must be as easily cleanable as possible (i.e., coved corners, smooth joints and seams, screws and bolts eliminated or buried).
3. Equipment must meet performance standards (i.e., a warewashing machine must clean a given number of items in a defined period of time).
4. Thermometers must be of an approved type, easy to read, and located in an accessible area.

Equipment Specifications

Equipment should comply with applicable UL, NSF, and American Gas Association standards, as well as fire, plumbing, and electrical codes that have been established by state and local governments. It is important that all connections for all equipment using utility service comply with OSHA regulations. Good specifications for a piece of food service equipment should describe accurately and exactly what is wanted. The information should include the following information:

1. The material desired for every part of equipment, such as 14-gauge stainless steel for the top of the counter with adequate insulation for the deadening of sound, and crossbars of the right strength for support, with 18-gauge stainless steel for the ends of the counter and galvanized steel for the back.

3. National Sanitation Foundation, P.O. Box 1468, Ann Arbor, MI 48106.

2. The voltage, phase, and cycles, such as 208, three-phase, or 60 cycles, should be specified if the item is electric, Btu's if the item burns gas.
3. The size of the item should be precise. This should include height, depth, width, and in some cases thickness.
4. The thickness of insulation should be considered in refrigerators, freezers, ovens, hot food serving tables, and mobile heated-dish storage units.
5. The number and location of thermometers, gauges, and thermostats should be checked for refrigerators, ovens, grills, fryers, braising pans, and dishwashers.
6. The size of the motor should be compared and specified on refrigerators, freezers, mixers, cutters, slicers, and dishwashers.
7. The size and location of compressors should be studied and specified for refrigerators and freezers.
8. The size and kind of casters must be given careful thought, as casters that are too small will not move easily, especially when the load is heavy, and casters that are too big may increase the cost unnecessarily. If wheels are to be locked for any purpose, locks must be specified. If carts are to be pushed, the type and location of the handles are important considerations.
9. Give the model number or numbers if available from more than one company describing the exact piece of equipment needed for your facility.
10. Requirements for delivery, installation, certification, and cleanup should be spelled out when the order is placed.

Most equipment companies offer a number of optional extras on equipment. Some of these options are necessary or at least make the piece of equipment more versatile, but some may be more in the line of gadgets. The inclusion of the extras will add to the cost of the basic item.

Carefully prepared specifications are necessary in preparation for bids. A bid is a quoted price for a specific item or items that are to be purchased. If specifications are so strict that only one company can bid on the item, the price may be higher than if the specifications were broad enough to allow several companies to bid. If brand names are used in conjunction with the "or equal" clause, substitutions need not be accepted unless they are exactly equal.

Leasing

If equipment is needed in order to improve efficiency, reduce labor costs, handle a larger number of patrons, or some other reason and funds are not available, the facility might consider leasing the equipment. Leasing provides an option for obtaining the item or items for immediate use when money is not available for purchasing it. A detailed contract specifying the responsibility for care and maintenance, right to purchase at a later date, and legal ownership should be prepared and signed by all parties involved in the lease agreement.

Some advantages of leasing equipment:

1. Equipment can be obtained though there are not enough funds for purchasing the equipment.
2. Rates are known, which facilitates budgeting.
3. Equipment might be replaced with newer models as design changes take place.
4. Leasing costs can be deducted as an expenditure before taxes are calculated.

Some disadvantages of leasing equipment:

1. The interest on payments for the equipment increases the overall cost of the equipment if it were purchased.
2. Leasing companies usually retain ownership of the equipment; the remaining value of the equipment when it is returned accrues to the leasing company.

3

Receiving
and Storing Food

Correct and efficient receiving and storing of food is the first requirement for a well-managed, profitable food service operation. The right equipment correctly placed and properly used will make this management function efficient and easy to perform.

Entrance Hall

The outside entrance should open into a vestibule or entrance hall, not directly into the food service area. This arrangement will prevent mud, flies, and other insects from getting into the food production areas.

Receiving Area

There should be a room or an area just off the corridor near the door where food products can be placed by the delivery man. If this area is planned efficiently, reasonable increases of deliveries will not affect its functioning capability.

There should be enough space for two or more people to maneuver carts and/or trucks at one time. A minimum space of 8 × 8 ft would provide room for a receiving table, a scale, and a dolly or cart. Since 8 ft is the standard width for a corridor, an uncluttered corridor could serve as the receiving area.

This area should equal 0.64 ft² per patient bed for a 100-bed health care facility, or 0.42 ft² for a 150-bed facility or a 300- to 450-meal school lunch program or similarly sized restaurant. A larger facility would need additional space for carts and/or for stacking items that have been delivered.

TABLES AND CARTS. A receiving table and a flatbed truck or sturdy utility cart will be needed in the receiving area. The table should be about 2½ × 4–6 ft; the smallest flatbed truck available is approximately 24 × 36 in., a handy size to operate. A larger truck may be needed in a large facility, such as one serving more than 500 meals per day. A security cart with the top and three sides made of heavy wire and doors on the front side that can be locked may be needed for supplies for use the next day that do not need refrigeration.

SCALES. Provide adequate space for good quality scales on the receiving table or use platform scales that stand on the floor. Inexpensive, lightweight, spring-operated scales may quickly decrease in accuracy through wear. The mechanism in the more expensive, heavy-duty scales is a beam that does not lose accuracy with use. An optional feature is the tare adjustment that allows the scale to be balanced with an empty container on the platform; thus weighing always begins at zero. This feature eliminates the need for subtracting the weight of the container from the total weight registered on the fan.

New electronic scales feature an electronic strain-gauge transducer (photoelectric cells) mechanism, which increases accuracy; there are no mechanical weights or springs. This type of scale has a capacity of up to 25 lb with a tare weight adjustment of up to 10 lb. The readout may be wall or table mounted. This unit requires electric power.

If meat is to be purchased as either portion-ready or wholesale cuts, a table scale with a capacity of up to 50 or 60 lb should be adequate. Scales of this size are available with 1-oz divisions, which are good for accurately checking the weight of the food being delivered. The size of the tray or platform should be compatible with the size of the packages to be weighed.

The United States is gradually adopting the international metric system (Metric-SI for Système International d'Unites), the standard of weights and measures used by most of the world. Food service employees will find the changeover less complicated if they are provided with a good set of scales that shows both the metric system and the customary system (pounds and ounces). Beam and spring scales are available with both metric and customary systems marked on the chart. Electronic scales with a switch that allows the operator to choose which system is to be used may be purchased.

Storeroom

A storeroom 8 ft wide and of a length suited to planned needs adds to the efficiency of the operation when located adjacent to the kitchen, near the receiving area. If it is desirable to have the storeroom wider than 8 ft, increase the area by multiples of 7 ft to allow for the addition of a double row of 21-in.-wide shelving plus another 3½-ft aisle.

SHELVING. The storeroom should contain adequate adjustable shelves for open cases of food. The linear feet of shelf space needed can be calculated by totaling the number of kinds of fruits, vegetables, juices, and other foods to be kept on the shelves.

Several types of shelving can be used satisfactorily: solid, louvered, embossed, mesh, or slat. Solid shelves are a little less expensive than louvered or embossed shelves. The materials most commonly used are vinyl-coated steel, a variety of plastics, anodized aluminum, or stainless steel. However, one type of shelving is a combination of nickel/chrome with an epoxy coating. This triple combination presumably provides excellent protection against rust. Thicker plating is provided on the tops of shelves and in areas of greatest wear. Stainless steel shelves are almost three times as expensive as vinyl-coated steel and about twice as costly as aluminum. The shelves should be adjustable, preferably on 1- to 2-in. centers and strong enough to support the weight of the items to be stored (Fig. 3.1).

Shelving to be used for canned fruits, vegetables, and juices should be 20–21 in. deep and spaced 17–18 in. apart

Fig. 3.1. Rack with adjustable shelves.

so that No. 10 cans, which are 7½ in. high and 6¼ in. in diameter, can be placed three deep and two high on each shelf. Two cases of one item in cans this size would require about one linear foot of shelf space.

Approximately 12–15 varieties and forms of fruits, 10–12 varieties and forms of vegetables, and 6–8 varieties of juice can be kept in an easily accessible storage area. Allow space for canned meats, canned instant potatoes, and other foods that might be available in No. 10 or similar size cans. Lightweight items such as prepared cereals can be stored on the top shelf.

Group food items in alphabetical order according to main categories and then arrange them on the shelves in an orderly manner according to size. Place small items, such as cans of baby foods, so that they are easily accessible; they may be kept in their original cases on very shallow shelves (not more than 8 in. apart) to save space and avoid breakage. More space between shelves may be needed for junior foods or No. 2½ cans of strained foods.

Store heavy products such as pickles and mayonnaise, especially those packed in glass jars or in heavy, odd-shaped cans, on the lower shelves. Gallon jars are usually about 11 in. high; therefore, the distance between shelves for those items need not be more than 12 in.

Many products packaged in odd-size cans and boxes may need to be stored near the preparation area. The important factor is to provide shelving that is adjustable on 1- to 2-in. centers and in widths and lengths most suited to your needs.

It is essential to allow for necessary but not excessive space between shelves. An excess of space between shelves could cause an increase of as much as 30% in the storage space.

High-density shelving provides maximum storage capacity in minimum space (Fig. 3.2A). These shelves are mounted on ball bearing rollers that glide easily on either floor-mounted or overhead extruded aluminum tracks, which have high-carbon steel liners (Fig. 3.2B). Since the shelves can be moved easily, multiple aisles can be eliminated, thus making room for even more shelves. Care must be taken to keep the tracks clean; the overhead tracks may be easier to keep clean.

Other features relative to shelves, such as back and end ledges, label holders, adjustable wall brackets, heavy duty casters, rubber bumpers, dividers, adjustable cushion feet, and/or adjustable ball feet, must be specified as additional items. These cost extra but may be needed in particular situations.

OTHER FACTORS AFFECTING STORAGE SPACE. Provide space in the large central storage area for extra cases of staples, canned foods, and paper supplies that the facility will need. Slatted floor racks are needed for supplies not stacked on shelves. The racks protect foods from dampness that may be absorbed from floors, permit the air to circulate freely around the foods, and help to keep the storeroom neat and orderly. A 3 × 4 ft rack is small enough to be easily moved for cleaning, yet large enough to hold up to 35 cases of canned goods.

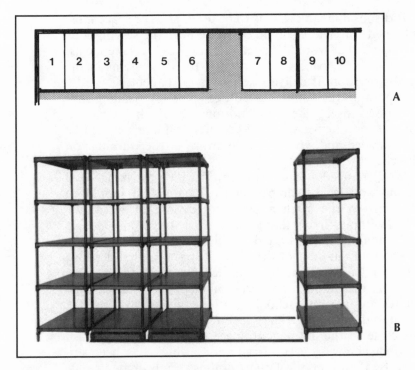

Fig. 3.2. High-density storage. **A.** High-density utilization of space; **B.** High-density shelving on ball bearing rollers on a floor-mounted track. (*Courtesy Market Forge, Everett, Mass.*)

If the central storeroom is on a different level from the food service department, a lift or an elevator must be available for transferring heavy loads of stored items from one floor to another. If the facility is located in an area where deliveries are made infrequently, additional walk-in or reach-in refrigeration and freezer space may be needed in the storage area.

Increasing use of frozen and other convenience foods in modern institutional food service has not greatly changed the estimated amount of dry-storage space. Packaged cake and biscuit mixes do not take up less space than bags of flour, and there are many new items such as dehydrated or freeze-dried foods that require shelf space. School lunch units need a separate area for commodities.

If disposable tableware is to be used, storage space must be planned for. There should be space near the serving area for a supply that would be adequate for at least three meals. A great deal of space will be needed for these items in the central storeroom if delivery service is infrequent.

Separate, well-ventilated storage areas for cleaning supplies, tools, insecticides, and poisons should be separated from any food or food-related supplies. All items should be clearly labeled and workers made aware of the warnings on all labels.

Workers need a separate room in which to change into uniforms and keep their personal belongings. It should be convenient to the kitchen but located so that it can be reached without first entering the kitchen. Individual lockers for personal belongings are desirable.

The total amount of dry-storage space needed depends on the purchasing methods used, the frequency of deliveries, as well as the amount and type of disposables used. Fast food restaurants and schools that use disposables need a large amount of space. The space for the disposables could come close to the amount of space needed for a dishwash area if dishes were used. Irradiation, which has been approved by the Food and Drug Administration, may also affect the kind and amount of food storage space needed in the future.

Space Determination and Layout: Food Preparation Area

The basic principles of a good work-flow pattern, adequate aisle space, and properly selected equipment apply to all food production situations and help make the best use of equipment available. The receiving and storage area should be accessible to one end of the preparation area with unencumbered flow lines from the production area to the serving area.

The simplest arrangement for an efficient food preparation area includes food production equipment, a cook's table, a baker's table, a refrigerator, and a pot-and-pan sink. These items would fit into a space 14 × 20 ft, or a total of 280 ft², allowing for 3½- to 4-ft aisles between tables and cooking equipment. Additional aisle space would cause the worker to take unnecessary steps. Oven, steamer, and refrigerator doors should not open opposite each other because this requires a wider aisle and therefore more floor space.

A workable arrangement for a food preparation area in a larger facility might include a bank of refrigeration (2½ ft deep), a work aisle (4 ft wide), work tables (each 2½ ft wide), 2 aisles (each 3½ ft wide), a double bank of cooking equipment (about 7 ft wide), and utility tables (each 2½ ft wide), which may be mobile or stationary (Fig. 4.1). Refrigeration might also be needed on the second side of the cooking island. The space for equipment and aisles totals about 25 ft in width. This work area would amount to some 20 ft in length if 2½ ft were allowed for a pot-and-pan washing area, 3½ ft for an aisle, 9 ft for cooking equipment, and 5–6 ft for

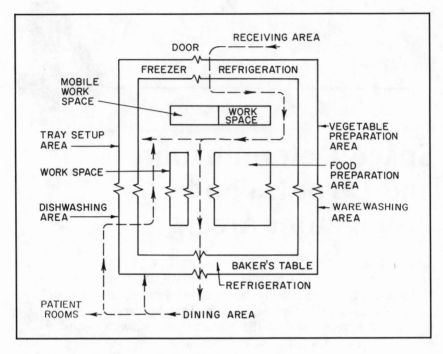

Fig. 4.1. Possible layout for food preparation area.

a traffic aisle. The area for this production arrangement would be 25 × 20 ft, or a total of 500 ft². This amount of space and equipment could easily serve 300–600 meals per day, depending on whether they were served at one service or three times per day.

The number of meals to be prepared can be greatly increased by the addition of steam equipment, a convection oven, a tilting braising pan, and adequate refrigeration. The addition of one or two pieces of cooking equipment requires very little extra space, but can increase the production capability a great deal. In many instances it is convenient to have the warewashing and vegetable preparation units located at right angles to the cooking equipment. If a larger warewashing machine is needed more space would be necessary; in fact a double sink, drainboards, and the machine might be located in a separate area or room. It is also important to provide space for utility carts in this area.

The food production area for much larger facilities may only be 30–32 ft wide but much longer than the area previously described. A bank of walk-in refrigerators and freezers plus a large day-storage area add to the width of a large facility; however, the basic space relationships are the same. The production area in a large school that serves 600–800 lunches is designed to accommodate a large amount of baked goods. The storage space would be large to accommodate the storage of disposables and commodities. The production area for a 350-bed hospital might allow for the preparation of many fresh vegetables and potatoes.

A hotel production area might differ by including a large section of broilers, grills, smaller steam-jacketed kettles, roast-hold ovens, and perhaps equipment for the cook-chill process in order to take pressure off the cooks when large banquets are scheduled.

5

Food Preparation Equipment

Technology has moved commercial cooking into the forefront of the modern era. Many pieces of equipment are now multipurpose, which not only saves space by reducing the number of pieces needed, but also improves the flexibility of utilization of a given area. Insulation materials are improved, temperature controls are improved by use of sensitive thermostats that hold temperatures accurately, and timers are now integral parts of much equipment, even to the point of automatically stopping the cooking process. Even the use of improved casters on more kinds of equipment has improved efficiency in the food production area. There are many improvements in the method of heat exchange in the cooking medium, which can improve efficiency by moving surrounding air or by the use of steam pressure, making it possible to get more controlled heat into the product in a shorter period of time. This is accomplished by use of such modern equipment as convection ovens, convection steam equipment, microwave ovens, pressure frying equipment, and high-pressure steam equipment.

Ovens

Ovens provide the principal method of volume dry heat cookery. Although the source of power is either gas or electricity, the method of heat supply may be conventional, convection, or microwave. Any oven that has passed the American Gas Association testing is considered adequately insulated. All cooking equipment should have National Sanitation Foundation (NSF) approval.

CONVENTIONAL. Conventional ovens are arranged either as decks or under a range top. Deck ovens may be single units or stacked two, three, or four decks high, but if decks are stacked too high, the top one will be difficult to reach and the bottom one too low for convenience.

Ovens should be insulated and completely sealed so that moisture will not get to the insulation material. Most conventional ovens have solid doors, so if glass doors are preferred, they must be specified.

Ease of cleaning is important to consider when selecting an oven. Porcelain enamel linings can be cleaned easily. Removable linings can be put through the dishwasher, but the linings can warp and become difficult to fit back into place. Stainless steel or aluminized steel interiors are more durable but not as easy to clean.

After an oven has been installed, check to be sure the thermostat is functioning properly. Ovens can be as much as 50°F (30°C) over or under the thermostatic setting. Some companies recommend that thermostatic controls be recessed and located at the side of the oven rather than above the door so that they will not be affected by seepage of hot air or by opening the door. Precise heat control is maintained by a fully insulated and vented thermostat. This added protection virtually eliminates the need for thermostat replacement and gives more years of accurate temperature control. All mechanical parts of the oven and thermostats should be accessible from the front for servicing.

ROTARY. A rotary oven combines the advantages of a modified convection system and the convenience of loading and unloading at a convenient height. Rather than having fans circulate the air within the oven, a wheel mechanism moves the product through the heated chamber. A single door provides access to all the shelves. Each shelf is advanced until it is in position at the door, allowing foods to be easily put in or taken out.

Ovens need to be selected on the basis of volume need. They range in size from 3⅔ × 6¾ ft, which hold 6 bun pans, to 17½ × 11¾ ft, which hold 80 pans. These ovens are great for facilities where large amounts of food need to be baked.

CONVECTION. Food in a convection oven is surrounded by heated air that contacts the surface. The heat is conducted through the surface and into the food until the food reaches the proper degree of doneness (Fig. 5.1).

The convection oven has a fan that constantly circulates heated air across the food, stripping away the cold layers around the food and replacing it with moving heated air. This convection action accelerates the heat absorption

Fig. 5.1. High-efficiency convection ovens, stacked. (*Courtesy Market Forge, Everett, Mass.*)

process, shortens required cooking time, and reduces shrinkage for conventional foods of the same density.

Heavy-duty electric or gas convection ovens require about 38 × 38 in. floor space. It is generally considered that one convection oven will produce as much as three single conventional ovens under range tops. These ovens can be stacked to provide a tremendous amount of oven space in a small amount of floor space, but this makes the top shelves very high and the bottom shelves very low, making access difficult. There is a convection oven 14⅜ × 20⅞ in. designed to accommodate 13 × 18 in. (half-size) pans. An oversized unit especially designed to accommodate hotel roasting pans is available also. There is one convection oven that can be installed under a choice of range tops.

Newer models of convection ovens are available with a roast-hold feature, which includes a 2-speed ¾-hp fan motor and a 60-min and a 5-hr timer. These ovens can be preset at 160°F (71°C) for holding roast meats. There is also a function switch to select either the roast-hold or normal operation of the oven. The roast-hold feature provides for longer, slower cooking, which reduces meat shrinkage but may not be desirable for ovens that are to be used for a lot of baking.

Load control lets you dial the amount of heat input to match the size of the load put into the oven. This feature is designed so that you set the dial on 100% for heavy load capacity, 60% for medium capacity, and 30–40% for low capacity.

The changes that are rapidly taking place in convection ovens include the use of dual air flow, tunnel air flow, and the use of heat exchangers behind perforated panels, all of which cause the heated air to be recirculated before being vented into the exhaust system, thus reducing the British thermal units (Btu's) from 100,000 to 60,000 or the kilowatts (kw) from 16 to 10 in many instances.

The circulated air is distributed evenly over the food product from the front to the back of the oven. Controlled air flow means that food is baked or roasted more evenly and efficiently. When the air moves to the back of the oven it does not blow out of the oven when the door is opened. The cooking time is reduced from 30% to 50% over that of conventional ovens.

Controls are solid state; therefore, there are no movable

parts to repair. Most control panels are removable from the front for easy access to parts such as the gas pressure regulator and the gas cut-off valve. Computerized controls, which include self-diagnostic checks and memory backup, are available.

Some convection ovens are equipped with thermocouple units, which are inserted into foods to give interior temperature readings for doneness. A light or buzzer signals when the proper interior temperature has been reached.

A good oven should rise to 450°F (230°C) within 20 min and should have good recoverability. The thickness of the insulation is not always as important as its effectiveness. The "K" value, or the amount of heat that goes through 1 ft² of 1-in.-thick material when there is a difference of 1°F, is important. Doors should be insulated also.

Some of the standard features for convection ovens are 2-speed fans, magnetic circuit breakers, adjustable legs, doors that open to a 180° angle to provide maximum access, five racks, and a pilotless ignitor for gas ovens.

Optional features are available but vary with brand and include casters (two swivel and two with brakes), additional timers (5-hr), solid-state meat probes, windows in both doors, additional racks, continuous-clean (catalytic process), spill-over pans that can be removed for easy cleaning, and a cabinet base with rack glides and cook-hold controls.

COOK-HOLD. This type of oven is only for cook-hold cooking. These ovens circulate heated air through baffles and ducts in the back and on the sides to provide even heat distribution throughout the entire cabinet. Venting is not required; therefore this unit might be used in situations where additional venting would be difficult. The oven cavity has a closed system of air circulation to ensure uniform roasting. All parts are removable for easy service and cleaning. The units can be stacked two high and are available on casters.

CONVEYOR. The heated air in conveyor ovens is directed to impellers that push the air through rows of fingers extending across the conveyor, applying heat transfer to the product (Fig. 5.2). The sweep action of the conveyor provides uniform rapid· baking from both top and bottom. Electronic temperature control adjusts the heat input to match the product load. The conveyor speed can be preset by a micro-

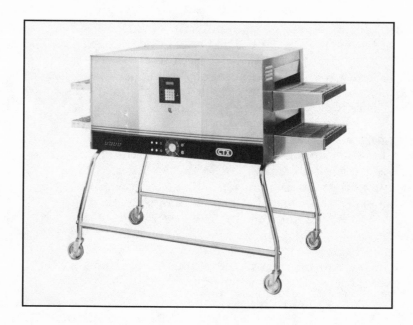

Fig. 5.2. Conveyor oven. (*Courtesy CTX Gemini, Fenton, Mo.*)

processor control. There is an automatic cool-down feature when the oven is turned off. The oven is built on a heavy-duty tubular steel frame and is well insulated for minimal heat loss. It was developed originally for baking pizza but is now being used for many other baked products.

MICROWAVE. Commercial microwave ovens are designed to rapidly heat or cook foods. In the microwave oven heat is not generated until the waves are absorbed by the food in the chamber. Different foods heat at different rates because of different densities (and shapes). The degree of doneness is determined by the length of time the food is in the oven and the power level used; most of these ovens are available with two or more power levels. Microwave cooking will not brown foods. However, manufacturers have indicated that calrod elements can be used for radiant exposure to compensate for this lack of browning.

There is no thermostat to set and there is no preheating of microwave ovens. The development of electronic control has permitted streamlining by combining time, power level, and start functions into a single program activated by a pushbutton switch. Decareau has concluded that an overall mean of about 75% less energy is required to cook or heat foods by microwave than by conventional methods.[1]

Considerable design knowledge is available today in "choking" microwave energy, which is defined as the effective elimination of microwave energy from areas where it might cause a problem. This is done by forming the metal components to known shapes in order to create a low-voltage region, thus preventing arcing.[2] Decareau has reported the considerations for wise use of metals in food service and has stated that in the future, metal will not be limited to shielding but rather will be used in containers and packaging.[3] Government standards have been set for the amount of energy leakage permitted.

Tunnel microwave systems have been developed and are being used in some cook-chill situations.

Steam Cooking Equipment

The use of steam equipment reduces the number of pots and pans to be washed, speeds up cooking, saves energy, thus reducing fuel bills. Steam cooking is adaptable to initial preparation, reheating, and reconstituting food. The food does not burn on the pan, so there is no scouring before washing. It also eliminates the lifting of heavy stockpots on and off the range top.

STEAM-JACKETED KETTLES. Size is an important factor in steam-jacketed kettles. If the kettle is too large, it is too difficult to use; if it is too small, extra batches must be made. Kettles are available in sizes ranging from 10 qt to 150 gal. A 15-gal kettle filled to 6 in. from the rim provides about 260 6-oz portions and 350 4-oz portions. This should be ample for a facility that may serve as many as 200 meals at a time.

---

[1]Decareau, R. V. "Developing Food Products for the Microwave Oven Market." *Microwave Energy Application Newsletter* 7(January 1975).

[2]Bowen, R. "Contemporary Views on Using Metal in Microwave Ovens." *Microwave World* 7(January 1982).

[3]Decareau, R. V. "Metal and Microwaves." *Microwave Energy Application Newsletter* 10(February 1976).

A 20-gal kettle filled to 6 in. from the rim furnishes about 380 6-oz portions, or 500 4-oz portions. The 10-gal unit is a good size for a facility that serves from 100 to 150 meals at one time.

The 20- or 40-qt models can be mounted on tables or stands no more than 30 in. high. Kettles mounted on tables or stands may have stainless steel dump drawers or a trough (which may be fitted with a permanent drain connection) to provide safe, easy dumping of product or cleaning water. This arrangement prevents the splashing that occurs when the kettle and the floor are not properly aligned. Single units of 20-, 30-, or 40-gal capacities can be mounted on a pedestal base, legs, or on a modular base that encloses plumbing connections and the drain (Figs. 5.3A, B, C). Drains must be located correctly and be large enough to prevent backflow when the kettle is being emptied. Adequate work space should be provided around the kettle for efficient food preparation and cleaning. A water source with a swivel faucet

A   B   C

Fig. 5.3. Steam-jacketed kettles. **A.** Self-contained, pedestal-based 40-gal kettle: ⅔ jacket, internal electric element, steam generated; **B.** Self-contained, leg-mounted 40-gal stationary kettle: ⅔ jacket, internal electric element, steam generated; **C.** Self-contained, modular-based 40-gal tilting kettle: pan support, boiler in base. (*Courtesy Market Forge, Everett, Mass.*)

should be available to provide water for food preparation as well as cleaning.

The more sophisticated equipment for special installations includes several types of wall-mounted kettles, including a cooker/mixer version. The floor can be cleaned with ease under these units, but it is sometimes difficult to reach behind these kettles to clean. A great deal of construction work is required to relocate a wall-mounted kettle. The cost of the L-shaped wall bracket used for mounting each kettle is several hundred dollars and the cost of installation is almost as much.

The platform console can be used to accommodate mounting any kettle combination. The complete assembly is available with legs for a floor-standing unit, or the console unit may be wall-mounted. The stainless steel platform features a built-in drain trough and a sliding dump drawer that can be slid under the drain area of any kettle. This provides an easy-to-lift container for safe dumping of product or cleaning water. The built-in drain trough can be attached directly to a permanent floor drain. Matching modular consoles house all steam and water connections, condensate-return piping, trunnion bearings, and self-locking worm and gear mechanisms. This type of unit eliminates the problem of cleaning exposed pipes or having a poorly located drain that may also cause cleaning problems.

Tilting, Trunnion, or Stationary Kettles. Small steam-jacketed kettles usually are the tilting type and do not have draw-off valves (Fig. 5.4). Large tilting kettles usually have draw-off valves, preferably the tangent type. Some older models may have perpendicular valves, but most new equipment is designed with a tangent valve. Tangent draw-off valves provide for the food to be removed along a straight horizontal path rather than having a right angle joint in which food can become lodged.

Remember to plan for a place to put the kettle contents, whether the place is a pan holder attachment or a low cart that can be drawn up to the kettle. This arrangement is necessary for removing food from large kettles. Trunnion kettles (those with a wheel-tilting device) allow food to be dumped or poured into the serving pan. This eliminates the necessity of ladling large amounts of heavy food from the kettle into

Fig. 5.4. Tilting 10-gal steam-jacketed kettle: self-contained, electric. (*Courtesy Market Forge, Everett, Mass.*)

serving containers. However, items prepared in a kettle are often a combination of solids and liquids, and these should be ladled out to get an even ratio of solids to liquids.

Full- or Two-thirds–Jacketed Kettles. A choice must be made between the two-thirds–jacketed and the full-jacketed shallow kettles. The two-thirds–jacketed kettles provide more capacity for the amount of floor space and reduce scorching around the top edges of the product because the product level is usually above the jacket section. Full-jacketed shallow kettles are better for browning and roasting meat.

Electrical, Gas, or Steam Connections. One self-contained electric steam-jacketed kettle for table-top use plugs into an outlet like an electrical appliance. A thermostat controls temperature ranges from 150°to 270°F (65–130°C). As the steam pressure in the jacket increases, the temperature increases in direct proportion. For each pound of pressure increase, there is a 3°F (2°C) temperature increase. At 15 lb of pressure, the temperature is 257°F (125°C).

Steam-jacketed kettles must be connected correctly so that the steam supply will be at the right pressure. If the steam pressure is above 15 psi, request a pressure-reducing valve.

A filter must be installed in the steam line to prevent foreign material from collecting and interfering with the steam supply unless clean, nontoxic steam can be provided. Ask the mechanical engineer to sign an agreement stating that clean, nontoxic steam will be available or that the proper type of filter will be provided.

If the available steam is wet steam, it contains too much water and will not get hot enough. This problem can be eliminated by putting ball-float traps in the steam line.

Water that accumulates in the jacket when the steam condenses is drained by means of the small valve on the bottom of the kettle. If this is not done before filling the jacket of the kettle, the cold water will prevent the steam from filling the jacket and cooking will take place only on the sides and not on the bottom. Water in the jacket often causes a loud rumbling noise.

Shutoff valves should be installed in the gas, water, and direct-steam lines. If a valve is electrically powered, locate a switch near the kettle. Boilers should have a low-water cutoff valve and a magnesium descaler.

Specify the size (gallons), type (pedestal, legs, or modular), tilting mechanism, and utilities (electric, gas, or steam) that will best serve the facility with the funds available. Directly connected steam units are much less expensive than the self-contained electric or gas units; however, clean steam is not always available. The tilting mechanism may add as much as 30% to the cost of the unit, but it is a valuable asset.

There are many optional accessories for use depending on size of facility, menus served, and resources. Scraper-type mixers are available for large steam-jacketed kettles, but they increase the cost three to four times. Units that are attached to the kettle are heavy to handle and not easy to clean. There is a portable mixer unit available that will fit the 20- to 80-gal-deep kettles. The combination rotary and lifting action provides good blending for stews, chili, spaghetti sauce, and soups; it also does an excellent job of mixing puddings and pie fillings. This portable unit can be securely attached to the kettle rim, and the lightweight mixer head is easily lifted in and out of the kettle. There is a sturdy support base that can be used to move the agitator against or away from the kettle or kettles.

Covers and baskets for various sizes of kettles cost extra. Special long-handled paddles and whips are both a convenience and a safety factor, because they allow for thorough mixing without requiring the operator to be close enough to be burned. The choices have to be made in relation to the needs of the facility and the budget.

COMPARTMENT STEAMERS. Compartment steamers are used in high-volume kitchens with short serving periods (Fig. 5.5A–D). The large double units can hold 6 (Fig. 5.5A) or 12 (Fig. 5.5B), 12 × 20 × 2½ in. pans; each compartment has separate controls. Some units can be used for cooking without pressure in the one compartment and with pressure in the other compartment. Other models can be converted at the flick of a switch from steam pressure for bulk cooking to convection steam for defrosting foods and cooking vegetables. Each of the two compartments operates independently. This type of compartment steamer is 55 in. high × 36 in. wide × 33 in. deep (Fig. 5.5B).

The compartment steamer may be equipped with a self-generating gas or electric boiler or may be connected to direct steam or a regenerated (steam coil) boiler. The units that connect to the steam system in a building cost less.

There are countertop models that are 29 in. high × 24 in. wide × 33 in. deep (Fig. 5.5C). These units are available with a base or legs. This steamer can be operated with or without pressure with the flick of a switch. Some models have a 15-min timer for precision control plus a 60-min timer for continuous operation. Also, some models include a load-compensated timer, which does not start timing until the compartment is saturated with steam. The smaller units have a built-in generator on the side or bottom, therefore no boiler is necessary (Fig. 5.5D). Steam condensate is removed

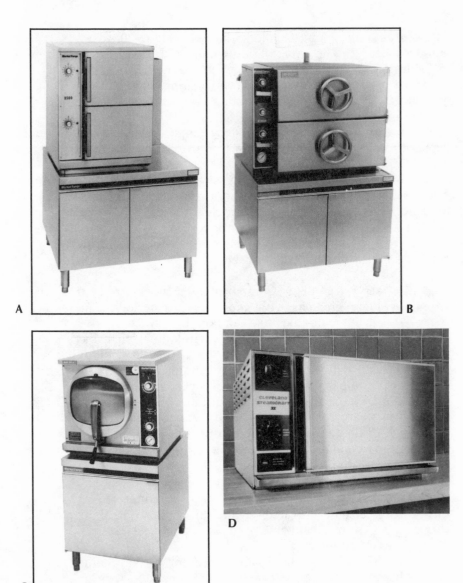

Fig. 5.5. Compartment steamers. **A.** High-volume, 2-compartment steam cooker/defroster; **B.** High-volume, 2-compartment steam cooker, can be converted to convection steam; **C.** Countertop steam cooker mounted on a base; **D.** Countertop steam cooker with built-in generator. (*Courtesy Market Forge, Everett, Mass.*)

and reheated, eliminating the water barrier that insulates the food; this allows for the most efficient means of heat transfer. Forced convection helps to remove the air and water variable for the cooking process.

Most models are insulated, seamless, and have coved corners. Steam automatically stops when the door is opened. Doors are fully insulated so that the exterior of the door is relatively cool and are made with a strong positive closing action.

The model number, size, and energy source (gas, electric, or direct steam) should be specified. Optional features include 15-min load-compensated timers and sliding shelves for pans smaller than 12 × 20 in.

Cook-Chill Equipment

Some facilities use steam-jacketed kettles and compartment steamers to cook food until tender crisp or as a finished product; pan the food, no more than 2 in. deep; then chill it immediately in a blast chiller until it drops to 38°F (3°C). More sophisticated systems process kettle-cooked foods, pump the product into plastic casings, then rapidly chill the sealed casings for refrigerated storage (see Fig. 5.16). The panned foods have a shelf life of 4–6 days and foods in plastic casings have a shelf life of up to 45 days.

This system is useful in large facilities that need to reduce labor in the production area and to function with a skeleton crew on weekends and holidays. The system is energy efficient for steam produces the energy needed for the steam-heated kettles and cook tanks (Fig. 5.6). Sanitary procedures are at the highest level since human hands do not touch the food throughout the cook-chill process when the process includes pumping the food into bags, which are sealed instantly. The system reduces the number of pots and pans to be washed, and rapid chilling retards bacterial growth. There continues to be a need, however, for equipment to prepare baked goods and some breakfast items.

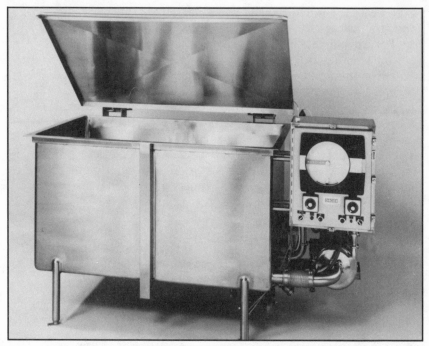

Fig. 5.6. Large cook tank used in cook-chill process. (*Courtesy Groen Div./Dover Corp., Elk Grove Village, Ill.*)

Other Cooking Equipment

Cooking equipment such as hot tops, griddle tops, broilers, and deep-fat fryers may be purchased as self-contained units, may be installed on a counter or base unit, or they may be part of a range unit containing an oven. The power source may be gas or electricity. If gas is to be used, specify whether it is to be natural or propane gas. Standard electric units heat more slowly than gas, so it may be desirable to specify one or more calrod units, which heat quickly, or choose gas as the source of energy.

HOT TOPS. Ranges may be purchased in modules 36 in. or 42 in. deep × 36 in. wide. It is possible to get any combination of tops by using modules of open-top burners, solid hot tops, or fry tops. These modules can be installed in different arrangements on counters, cabinets, or over refrigerated bases or ovens. Sectional ranges with ovens below may be the best choice when space is limited. A 2-section gas range with an oven below each section (some solid and some open top) can be set in a space 3 × 6 ft. A warming shelf above the top is useful for heating plates and covers. Front-fired hot tops allow a cook to boil in front and simmer in back. Sloped flues carry heat at diminishing temperatures toward the rear, giving gradient cooking temperatures.

DEEP-FAT FRYERS. Fryers are available either as cabinet (floor-mounted) or countertop models. It is wise to specify a fully insulated stainless steel tank with a large drain. If the unit is floor mounted, the fat is drained into a container placed below it. If it is a countertop model, the fat has to be drawn off through a petcock or siphon or has to be tilted and poured. A filter kit drains and strains the fat in a single operation. Some fryers provide a system of moving the oil continuously through a filtering system, which extends the life of the shortening or oil and also eliminates the need for a separate filtering system. There is also a filtering system that works like a vacuum cleaner whereby the oil is pulled out of the fryer, filtered, and pumped back into the fryer. The fry-saver filter removes oil-destroying contaminants. This system can be used in all fryers where the drains are located in the center of the fryer.

One of the most important specifications to consider is the kw or Btu rating, which indicates the recovery speed. The faster the recovery the less grease absorption by the product and the greater the productivity.

Fryers that do not filter continuously are designed to have a large cool zone under the elements for crumb accumulation and have less fat breakdown; they are also easier to clean.

Some fryers combine convection cooking, continuous filtration, and the use of heat exchangers for a more efficient performance.

The controls on fryers have been greatly improved: solid-state ignition eliminates all moving parts; burners cycle on and off, providing a holding temperature with less than a 50°F (25°C) swing; melt-cycle switches allow the melting of hard shortening. The thermal efficiency of fryers has been increased from 30% to 75%.

On/off switches and precision temperature-control dials and timers can be on the back panel or on the front of the fryer. Some fryers also have lights to tell when the power is turned on, when it is heating, and when the fat is up to cooking temperature; a few tell when the thermostat is not operating properly. Fryers should have a circuit-breaking device in addition to the regular thermostat to shut off the unit if the thermostat malfunctions.

Baskets available include a single large basket or twin baskets, each of which uses half the fryer. It is good to have a hook on the back of the fryer so that the basket can be suspended over the fryer to allow the food to drain. Handles usually project over the front of the fryer.

Several fryers have timer-actuated basket lift-out devices that will lower the basket into the fat for a prescribed period of time and then raise it out automatically (Fig. 5.7). Fryers are available with easy-to-operate, computerized digital read-out controls. The control package may offer a melt cycle, 2-stage temperature settings, a timer, and safety devices that shut the fryer down if the temperature gets too high. Some units have been designed to include an automatic, self-contained intermittent filtering system.

Fig. 5.7. Deep-fat fryer with timer-actuated basket lift-out device. (*Courtesy Vulcan-Hart Corp., Louisville, Ky.*)

Optional features include automatic basket lifts, melt-cycle controls, frypot covers, filter systems, flue deflectors, and computerized control systems.

TILTING BRAISING PAN (SKILLET). The tilting braising pan is one of the most versatile pieces of equipment (Fig. 5.8). This pan will thaw frozen food; handle griddle work; deep fry chicken, fish, etc.; stew and/or simmer; serve as a Bain Marie (hold food hot by setting pans or kettles of food in water); serve as a steamer (pressureless atmospheric cooking for shallow steam table pans set in water while lid is closed); and serve as an oven for baking and for roasting meat at low temperatures. A great deal of heavy lifting and transferring of foods from one pan to another can be eliminated with the use of a tilting braising pan. For example, stew meat can be browned and vegetables added for cooking all in one pan. If the tilting braising pan is to also take the place of a compartment steamer, it has to be large enough for the proper number of 12 × 20 in. pans to be used.

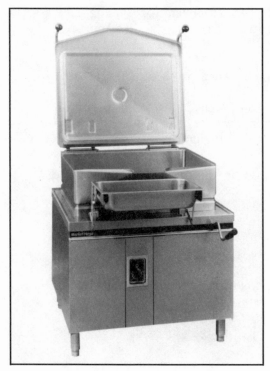

Fig. 5.8. Tilting braising pan with lid raised. (*Courtesy Market Forge, Everett, Mass.*)

The temperature in a tilting skillet ranges from 100°F (40°C) to 450°F (235°C). Most tilting skillets have an automatic thermostat for temperature control, a 60-min timer, a pouring lip, and positive control at any tilt angle. The size of these skillets or braising pans ranges from 36 in. to 64 in. in length, 32 in. to 34 in. in width, and 7 in. to 9 in. in depth. The overall height of the equipment is from 64 in. to 72 in. with the lid raised; the working height is 34–36 in. There is a choice of leg-mounted or modular-based units in 10-, 17-, 23-, 30-, or 40-gal capacities.

Electric or gas chambers are designed to produce usable heat below and around the cooking cavity. The insulated construction transmits very little heat to the surrounding areas. The stainless steel lining is seamless and easy to clean; however, it is important to have a source of water nearby and a floor drain directly below the pouring lip. Gas-fired skillets are somewhat less expensive than electric units, and open-leg models are less expensive than modular or wall-mounted units.

Specify the model number, size, and type of utility (gas or electric) to be used.

Optional extras feature lift-off covers for some models, pan holders, spray hoses for wash down, and steamer pan assemblies.

GRIDDLES. Griddles can be installed on the counter or back bar, be mounted on wheeled stands, or be part of a range unit (Fig. 5.9). Griddles should deliver fast preheat and instant temperature recovery so that they can be loaded continuously with frozen foods without significantly lowering the temperature.

The thickness of the griddle top is related directly to the drop in temperature when loaded with food. A griddle top that is ½-in. thick weighs 20 lb/ft^2. When cold food is placed on the surface, the temperature drops several degrees very quickly but also recovers to the original temperature very quickly. A griddle top that is 1-in. thick weighs 40 lb/ft^2. When cold food is placed on the surface the temperature drops a few degrees very slowly, but recovery is also slower than with thinner grills. A smooth surface resists sticking of foods.

Heating elements clamped to the underside of the grid-

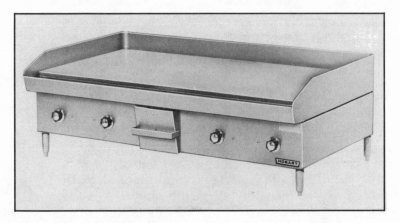

Fig. 5.9. Griddle with recessed controls. (*Courtesy Hobart Corp., Troy, Ohio*)

dle plate need high power input. Baffles mounted below heating elements force heat up to the griddle plate. Heating elements are patterned so that every square inch of griddle plate (including the perimeter) is uniformly heated to eliminate hot or cold spots. The automatic thermostat controls measure exactly the preset temperature. The exact location of the thermostat control should be indicated on the griddle top. It is important to place the first items to be cooked directly on the control, thereby triggering the power input to provide instant heat. The temperature range on most models is between 200° and 450°F (95° and 235°C). The griddles that are 36–72 in. in length have from two to eight independent controls, each measuring the heat for its section on the griddle plate. The advantage of this arrangement is being able to cook foods at different temperatures at the same time or to shut down parts of the griddle during the off-peak period.

Make sure a one-piece grease splash guard is welded continuously on three sides of the griddle plate to prevent grease leakage. The griddle should have a slight forward pitch toward the drain or grease trap, which is designed to lift out for easy cleaning. The grease receptacle needs to be large enough to handle all the drainoff that occurs during one period of usage.

A recessed control panel prevents accidental changing of dialed temperatures. Large signal lights indicate when heating units are energized. On some units, the lights shut off when the dialed temperature is reached.

BROILERS. Broilers are needed in facilities where tender cuts of meat are cooked to order.

Grids can usually be raised or lowered to several different positions. They should slide on rollers so that they can be pulled clear of the heating zone but not out of place. A grease drip shield that allows grease to drain into an easily removable grease pan is important for this unit.

When gas broilers are selected, be sure that burners and gas valves are located to make every grid position usable.

Some electric broilers have infrared rays that penetrate the product and reduce broiling time.

Broiler-Grill Combination. This unit uses snap-in quartz lamps for the top side of the unit or broiler. The controls are fixed so that three or five lamps can be operated at one time. The quartz lamps heat up instantly, thus reducing grill preheat time.

The broiler may be lowered to the cooking position since it is fitted with a spring-loaded mechanism.

The grill is fitted with a 7-qt grease bucket and removable liner. It is thermostatically controlled and absorbs unused heat from the broiler.

It is available in 304-18-8 nonmagnetic stainless steel with a no. 3 finish; 208 or 240 volts are necessary to supply the electrical needs of the unit. It must be vented by a normal hood.

Electric Char-Broilers. Electric char-broilers have tubular-type electric heating elements located under a grate area. The element assembly should pivot at the rear so that it can be swung up to facilitate complete removal of the cast-iron grate and drip pan for cleaning. Be sure elements are not more than ¾ in. below the top of the grate and designed so that the meat does not come in direct contact with the heating element. Grease and drip pans need to be easily accessible.

HEAVY-DUTY HOT PLATES. A 2-burner unit is frequently

needed by the cook and the baker for preparing or heating small amounts of food quickly (one egg, a bowl of soup, or a cup of melted butter). The burner on a gas unit should be designed to accommodate small utensils as well as large stockpots. All gas burners need pilot lights. Most gas burners have extra heavy-duty cast-iron grates (12 × 13 in.).

In an electrical unit a heavy-duty model will require 208–230 volts. Thermostats are necessary for heat control.

SPREADER PLATES. Additional work space and space for foods in process are needed between fryers, ranges, team-jacketed kettles, and steamers. Spreader plates should be well built and have a back splash if they are banked next to equipment that has a back splash. Fasten them securely to adjoining equipment with cantilever braces that will support heavy loads. Different sizes are available depending on need and available space.

Mixers, Cutters, and Slicers

Mixers are available in sizes ranging from the small 10-qt models to the 80-qt floor models (Fig. 5.10). Almost all these units can have a universal hub for attachments such as meat grinders or vegetable choppers so that the motor gets more efficient use.

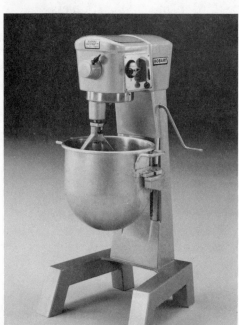

Fig. 5.10. Floor model mixer. (*Courtesy Hobart Corp., Troy, Ohio*)

BENCH AND FLOOR MIXERS. Mixers are available with a bowl capacity of 10, 12, 20, 30, 40, 60, and 80 qt. The size of mixer to buy depends on the greatest volume of food that will be needed at one time. The number of servings of mashed potatoes to be prepared is a good guide; 40–50 servings will about fill a 12-qt bowl. If a 12-qt unit is adequate, it may be better to purchase a mixer with a 20-qt capacity and a 12-qt bowl, particularly if there is a plan for expansion. A 20-qt mixer costs very little more than a 12-qt mixer. If you purchase the larger unit with a 12-qt bowl, the 20-qt bowl can be added later when volume increases.

A table model with a 20-qt capacity is about 16½ in. wide × 20 in. deep. It is usually placed on a low cabinet or table 24 in. high. There also should be a shelf or hooks for attachments. The table or cabinet may be purchased with or without casters; if the cabinet is on casters, there should be brakes to hold it firmly in place while the mixer is being used. A floor model with a 30-qt capacity is about 41 in. high and will occupy a floor space of about 20 in. × 20 in. Provisions should be made for storing attachments if a floor model is chosen. Different brands may have different horsepower on mixers of similar size.

Check equipment specifications to see what comes as standard equipment with the mixer. A bowl of heavy, tinned steel is usually included as a standard item. A stainless steel bowl has to be specified and will cost extra. Although a tinned bowl may cost only one-third as much as a stainless steel one, the tin wears off eventually and the bowl must usually be sent away for retinning. This means the mixer is inoperative while the bowl is being repaired unless an extra bowl is purchased for use during the repair process. Therefore, it is wise to consider specifying the stainless steel bowl when the mixer is purchased.

Mixers have become commonplace in most food service operations, but even the most experienced cooks should exercise caution to be sure the motor is off before attempting to scrape down the bowl and beater. They should also be sure the gear is set at the lowest speed until the ingredients in the bowl have been partially blended.

In addition to the bowl and flat beater, which usually come as standard equipment with a mixer, a variety of extra beaters and whips are available at extra cost. The wire whip is used to incorporate air into light mixtures. It is efficient for blending and whipping but should not be used in combining heavy mixtures. If heavy mixtures are beaten often, a 4- or 6-wing wire whip may be needed. A dough hook is used to mechanically knead doughs that require folding and stretching action. A pastry knife has one relatively sharp edge that cuts the fat into the dry ingredients.

There are several attachments available for a mixer when an attachment hub is purchased: vegetable slicer, grater, shredder, dicer, French-fry cutter, meat grinder, food chopper, coffee mill, fruit juice extractor, and knife sharpener. Be sure to specify a universal attachments hub so that more than one make of attachments can be used with it.

A speed-drive attachment increases the speed of grinders, cutters, etc., reducing the length of time needed for an operation. The speed-drive will add considerably to the cost of the mixer, so its value depends on how much the extra minutes of labor would cost and whether the operators are able to keep ingredients supplied to and removed from the machine at the necessary speed. Other attachments that may be of real value are the splash cover, bowl extension, bowl truck, pan bracket, and oil dropper.

Large mixers are available with three to five sizes of bowls; 80- and 140-qt models are available with a tall column design, allowing the removal of the bowl without removing the agitator. The bowl can set on the truck (wheel with casters), which can be rolled to where the product mix is to be used.

VERTICAL CUTTER/MIXERS. Vertical cutter/mixers can be used for cutting, mixing, blending, homogenizing, emulsifying, kneading, or pureeing (Fig. 5.11). They are available in 30- or 45-qt sizes. The bowl usually has a pouring lip and tilts for ease of emptying and cleaning. Adequate voltage and a floor drain are necessary at the point of use.

Standard attachments include a 6-min timer, cut/mix attachment, knead/mix attachment, strainer basket, mixing baffle, and cover scraper.

CUTTERS. Food cutters are machines that rapidly chop foods for salads, soups, dips, etc. (Fig. 5.12). The stainless knives rotate at high speed and save labor. The bowl covers and

Fig. 5.11. Vertical cutter/ mixer. (*Courtesy Hobart Corp., Troy, Ohio*)

Fig. 5.12. Food cutter with dicer attachment. (*Courtesy Hobart Corp., Troy, Ohio*)

stainless steel bowls are easily removed; therefore the machine is very easy to clean. The food cutters range from ½- to 1-hp motors and from 14- to 18-in. bowls.

SLICERS. Slicers are made of anodized aluminum and range from ⅕ hp to ½ hp with circular steel blades that are from 10 in. to 11¾ in. in diameter (Fig. 5.13). The thickness of a slice can range from paper thin to ¾ in. to 1¼ in. The smaller blade models are manually operated but feature a gravity feed. The models with a large blade are usually automatic with a choice of two speeds. Some of the new models can be used to slice hot or cold meats, vegetables, or even soft cheese.

Fig. 5.13. Compact food slicer. (*Courtesy Hobart Corp., Troy, Ohio*)

Refrigeration Systems

Guidelines for the general amount of refrigeration needed are meat and poultry, 35%; fresh fruits and vegetables, 30%; dairy products, 20%; frozen foods being thawed, 10%; and carry-over foods, 5%. The amount of refrigerator storage space needed is determined by the type of menu; the frequency of delivery; the amount of canned, frozen, and dehydrated foods used; and if the milk used is fresh, canned, or dehydrated and comes in bulk or individual cartons.

About 25–30 lb of food may be stored in 1 ft³ of space. This figure accounts for loosely packed items such as lettuce, as well as compact items such as butter. The net cubic-foot capacity represents only usable storage space.

COMPONENTS. It is important to select a well-insulated refrigerator that maintains the same temperature and humidity in all parts of the cabinet regardless of how full it is.

There are three major components in a refrigerator circuit: evaporator, compressor, and condenser. The evaporator consists of a series of coils surrounded by fins through which the refrigerant flows. Refrigerant starts out as a liquid but absorbs heat from inside the refrigerator and becomes a gas. The gas then moves from the evaporator coils into the compressor, a pumplike mechanism located on the exterior of the cabinet, which pumps or compresses the gas under pressure until it occupies a small area. From the compressor the gas goes into the condenser, which is also located on the exterior of the cabinet and adjacent to the compressor. The condenser consists of a series of coils surrounded by fins through which the compressed gas flows. There is a fan adjacent to the coils, and as the compressed gas flows through the coils the fan cools the gas. As the gas cools, it becomes liquid again. The liquid flows back into the evaporator coils under pressure and the cycle starts over.

The fins on the evaporator and the condenser must be kept free from all ice, dirt, and grease, which could block normal circulation of air and result in an overheated motor or complete stoppage of the refrigeration.

Most foods keep best at 33–40°F (1–5°C) and about 80% relative humidity. Relative humidity indicates how much of its total water-holding capacity the air is using at the moment. To achieve the desired humidity, the evaporator coils must be large enough to operate at just a few degrees lower than the cabinet temperature and still maintain adequate cooling capacity. This small temperature differential reduces the amount of condensation on the evaporator coils and maintains a higher moisture content in the cabinet.

Top-mounted refrigeration systems allow total interior cabinet space for food storage and handling. The condensing unit, condensate evaporator, coil housing with plasticized coil, and automatic defrost are mounted on top of the cabinet.

As more efficient insulating materials are developed, insulation that has been 2–2½ in. thick in most units is being reduced to 1 in.

WALK-IN REFRIGERATORS. Walk-in units need to be placed in a pit below floor level or set on screeds to provide about 3 in. of insulation below the refrigerator. If the unit is set on screeds, a 24- to 30-in. ramp will be needed to provide a gradual incline. Enough aisle space for the ramp and maneuvering space for the mobile equipment should be provided. Because of greater heat loads, these units must have adequate air exchange and ventilating systems.

A walk-in refrigerator is sized in square feet, and food is usually stored on mobile shelves that can be moved out for easy cleaning. The shelving usually varies from 18 in. to 24 in. in depth. Walk-in refrigerators should be approved by the NSF, be level with the floor, have tongue-and-groove joints, have a door latch that includes an interior safety release, have a thermometer outside the box that can be easily read, and be well insulated.

A walk-in refrigerator is a chilled storage area for cartons, boxes, and crates of food. Facilities that do not receive frequent deliveries, that use a large amount of fresh produce, and that buy in large quantity need walk-in refrigeration. These units are for the big, rough, storage jobs or for handling reserve products in school facilities where government surplus items are available and must be stored on short notice for extended periods if cost savings are to be realized. Walk-in refrigerators should be purchased for places with central commissary operations.

If a facility serves more than 300 meals per day there

may be a need for a walk-in unit in the receiving area for bulk storage. A larger facility will need a walk-in unit for each of the following: milk and other dairy products; fresh fruits and vegetables; meats; and poultry and fish. Also, a walk-in freezer will be needed. If a facility is using a cook-chill system, a holding unit (38°F, 3°C) will be needed.

REACH-IN REFRIGERATORS. Reach-in refrigeration allows for more flexibility in shelving and location (Fig. 5.14). Refrigeration should be provided in the receiving, preparation, and serving areas to increase efficiency of operation and reduce trips for supplies. If additional refrigeration is still needed, consider walk-in cold storage.

Under-the-counter units are convenient when installed near grills, broilers, and microwave ovens. This improves the efficiency of the operation by eliminating steps during the

Fig. 5.14. Reach-in refrigerator with food file, pan file, and rack file shelving. (*Courtesy Hobart Corp., Troy, Ohio*)

preparation process. Refrigerated drawers are usually available in under-the-counter units but may not be worth the extra cost unless they fill a specific need in a particular operation.

Stainless steel is the most durable material used in cabinet finishes, but it is also the most expensive. A stainless steel finish will cost about 20% more than a vinyl finish in a refrigerator of the same size. Porcelain and vinyl applied to steel and/or aluminum alloy have been declared acceptable materials in the NSF code. Liners should be completely free from seams. A 3-in. wall thickness of polyurethane foamed-in-place insulation is more than adequate for medium- and low-temperature units; many reach-in refrigerators are equipped with only 2 in. of insulation. It is important to specify the electrical characteristics as to voltage, frequency, and phase of the available current.

A warranty is usually issued by the manufacturer to cover defective parts and materials in the original product. If the manufacturer offers a 1-yr warranty on parts and labor, plus a 5-yr warranty on the compressor, you can assume the equipment will operate in accordance with the manufacturer's established standards.

The optional features provide variations of storage possibilities. Pull-out or roll-out shelves in refrigerators are for storing heavy items. Baskets that are 18 × 26 in. or 13 × 18 in. plus a lip to set on the ledge of the tray slide are in the receiving or preparation area where fresh fruits and vegetables will be stored.

Extra wire shelves are needed for refrigerators in preparation areas for holding jars and cans of fruits, salad dressings, or pans of cooked food. The distance between shelves will differ with the types of food to be stored.

Equip a reach-in refrigerator in the serving area with adjustable 1-in. food-file shelving centers or a roll-in cart to accommodate the 18 × 26 in. bun pans or wire shelving.

Factors determining the amount of reach-in space needed in this area:

1. The size of plates on which food is portioned.
2. The number of plates or items that fit on the 18 × 26 in. bun pan.
   a. Salad bowls — 11 per pan.

b. Fruit dishes—21 per pan.
c. Dessert plates (6½ in.)—11 per pan.
d. Juice glasses—84 per pan.
e. Milk cartons—40 per pan.
3. The amount of clearance between trays (the height of the food or beverage and the plate plus clearance).
4. The number of trays that can be stored per door-opening at the specified distance needed between trays.
5. The total number of portions to be stored at any one time.

Doors may be full height or half doors. Specify whether a right or left opening is desired. Locks are available for refrigerator doors but will cost extra. Pass-through units cost 20–25% more than units with hinged doors on the front of the unit. A 2-cell unit is less than 50% more than a 1-cell unit. A heavy-duty ball bearing roll-out shelf costs about twice as much as a pull-out shelf but is much more useful. The important thing is to find out what the optional extras actually cost and then select the things that will best meet the needs and yet stay within the budget.

ROLL-IN REFRIGERATORS. Front opening roll-in or pass-through refrigerators are ideal for intermediate storage between preparation and serving areas (Fig. 5.15). They can be used in receiving areas if properly designed carts are available. The carts need to be sturdy with heavy shelves that can be adjusted to hold cases of eggs, milk, lugs of fresh produce, and/or packages of meat. The carts, which could be moved to and from the receiving area, should be designed to fit the cavity of the cabinet.

Roll-in cabinets are for installation directly onto an existing level floor. Roll-in units may be level with the floor or may have a small entrance ramp. Some ramps are equipped with anticondensate heaters. Many roll-in units contain polyurethane plastic, which is poured, expanded, and bonded into place, giving good insulation in a relatively thin wall. The full-length doors are usually equipped with heavy-duty handles and cylinder locks that have an inside release safety mechanism. Interior lighting is available on units of this type. No drain is necessary because condensate is automatically evaporated. Roll-in units are available in one, two, or three sections.

Fig. 5.15. Refrigerator with roll-in cart. (*Courtesy Hobart Corp., Troy, Ohio*)

RAPID CHILL. Cook/rapid chill systems can reduce labor in the production area and can be very useful in situations when it is hard to cover weekend work or where food is prepared in a central area for satellites. The rapid chill system quickly cools precooked foods, reduces the time that bacteria grow well, and increases the storage life of food. The rapid chill system claims to be capable of cooling a product load to 45°F (7°C) in 1–2 hr. Two systems that are in use place the food, while it is still at its highest heat temperature, in shallow pans on an open rack, which is then rolled into a blast chiller. There is also a system that provides a rack carousel that permits rotation of the rack of food, thus cooling the food quickly and eliminating warm air pockets, which reduces chances for bacteria to grow. As soon as the food is cooled to 38°F (3°C), it is transferred to a holding refrigerator and kept at that temperature 4–6 days. All food is labeled and dated. Another system uses a tumbling chiller tank, or casing cooler. Food is sealed in Cryovac (for meat) or casings (for other foods) before being cooked in a cook tank and then cooled in the same tank or in the tumbling chiller (Fig. 5.16).

The tumbling chiller is a perforated drum that rotates in a tank of circulating chilled water. The casings are very strong and can withstand the mechanical kneading action of

Fig. 5.16. Tumbling chiller for cook/chill system. (*Courtesy Groen Div./Dover Corp., Elk Grove Village, Ill.*)

the rotating drum. The chilling time is carefully controlled, and the chilled product is transferred to storage for further chilling to 28–32°F (0°C). Foods handled in this manner can be held 30–45 days.

MILK DISPENSERS. Milk dispensers in which 3- or 5-gal cartons of milk are upended are commonly used in food service facilities. Units are available that store one, two, or three containers. The flow may be regulated for a certain portion each time, or it may be controlled by the person drawing the milk.

Specify whether the dispenser should have a front or back opening. The back opening is preferable if space allows, so that the machine can be loaded if necessary while in service. Some units also have a temperature indicator in the door making it possible to be sure the milk is kept between 32° and 36°F (0°and 2°C).

Another type of dispenser is a refrigerated self-leveling unit for cartons of milk. A dispenser of this type requires removal of all leftover milk cartons when filling the dispenser with cartons of fresh milk. Also, it is not always easy to know what percentage of cartons in each rack should be whole, skim, chocolate, or buttermilk.

Large milk storage units are available for storing cases of individual milk cartons as they are delivered from the dairy. These save the time of transferring the milk from the case to the rack in the self-leveling dispenser. There is also the advantage of being able to more easily shift the older milk so it can be used before the fresh milk. Be sure the size of the unit correlates with the case your dairy is using. These storage units are on heavy-duty swivel casters and can be moved to the delivery area then back to the serving area.

Freezing Systems

Freezing systems are designed to lower the temperature of foods to where they freeze solidly and can be stored over extended periods of time. The kind of food to be frozen, how it is to be prepared, the amount to be prepared, and how it is to be stored are considered in determining what system is to be used.

PLATE. This process involves direct contact between the packaged product and the freezing unit, which has been chilled by a refrigerant. This is the oldest and most widely used method.

RAPID BLAST AND RAPID FREEZE. Rapid blast and rapid freeze are systems that use fans on a fan panel mounted perpendicular to the cooling coils, which contain the refrigerant freon. The fans force high-velocity cold air over the food, stripping away the insulating layers of warmer air. This increases the rate of heat transfer. The freezer has a series of baffles so that the product gets maximum exposure. A rapid-blast freezer is designed to freeze large quantities of precooled food to $-35°F$ ($-37°C$) in a short time. A rapid-blast freezer can freeze 200 lb of precooled food in 8 hr, while a conventional freezer will take about 25 hr to freeze the same amount.

LIQUID NITROGEN. In the process of liquid nitrogen freezing, heat is absorbed, which turns the liquid into a gas. This vaporization of liquid nitrogen to gaseous nitrogen produces a temperature of $-320°F$ ($-195°C$), freezing the product very quickly, thus reducing bacterial growth that occurs during the cooling process.

LIQUID REFRIGERANT. This system does not involve exposure of the product to a gas or forced air. The product is immersed and has direct contact with the refrigerant at a temperature of $-35°F$ ($-37°C$). If freon is used, it is prechilled by being put through a mechanical refrigeration system until it reaches the proper temperature.

Some facilities use a system where the cooked food is placed on plates before being frozen. This system often creates a great deal of waste, particularly when patients on selective menus or restaurant patrons prefer to make their own selections.

LOW-TEMPERATURE HOLDING FREEZERS. These units are not designed to freeze food items but to store frozen foods at about 0°F (−18°C); therefore, the term freezer is really incorrect. Frozen food must not be allowed to stand at room temperature for any length of time; it must be stored at 0°F (−18°C) the minute it is received. As the temperature of frozen food rises above 0°F (−18°C), the rate of deterioration increases rapidly.

The insulation in frozen food storage units is usually 4 in. thick; the additional 1 in. (4 in. instead of 3 in.) is very important in the low-temperature units.

The compressor for a frozen food unit is usually twice as large as for a medium-temperature refrigerator of the same size.

ICE MAKERS. Before selecting an ice maker decide what kind of ice is needed. The three main kinds of ice are shaved or flaked, cubes, and chunklets. Shaved or flaked ice is made when a thin surface of ice forms on the inside of a cylinder and an auger system removes the ice from the cylinder. This is 40–60% true ice, has more surface area, less total cooling effect, and will dilute the beverage. Cubes are nearly clear, are available in a variety of configurations, and are produced by freezing the ice with a refrigerant in coils and then releasing the ice by introducing heat to melt the perimeter. Chunklets are made by first producing shaved ice then squeezing the water out to make a product that is 90% true ice.

The efficiency of all ice machines and the volume produced within a given time are closely related to the temperature of the room and the water being used. Increased temperature of either the room or water decreases production.

Polyurethane foam is one of the more effective insulations used in ice machines for it is space saving, rigid, and efficient. Fiberglass storage bins with contoured corners are durable and easy to clean. Thermostats should be at the top of the storage bin so that production automatically shuts off when ice builds up to its current capacity level.

Specify the size of the compressor and electrical characteristics. Water-cooled units make a larger volume of ice per hour than air-cooled units but are more expensive to operate. The production rate and the storage capacity will depend on the size and number of servings needed in a period of time. In general, 2 oz are considered to be a standard portion, but if water pitchers are being filled, a larger volume is required. Calculate needs on peak demand in relationship to storage capacity and hourly production. Avoid comparing the 24-hr production volume to the 24-hr usage because few if any facilities have constant demand.

Some facilities prefer to put an ice dispenser on the serving line and to locate the ice maker in an out-of-the-way space. This is particularly true in a facility where a large unit is needed or where ice is needed in several locations. A big advantage of an ice maker–dispenser combination is the elimination of unsanitary hand dipping or scooping of ice.

Thawing Systems

Systems for thawing are designed to raise and hold the temperature of frozen products at 40–45°F (5–7°C).

RAPID THAW. The rapid thaw refrigeration system uses high-velocity air flow and a special system of alternating cycles of heating and refrigeration. The air that circulates through the product zone never exceeds 45°F (7°C); as a result, the product never exceeds this temperature. There are visible controls to make sure the system works correctly. The thaw timer, which activates the food auxiliary fans, may be set for the length of time you expect it to take to thaw the product. When the timed cycle is completed, the fans automatically shut off and the cabinet reverts to a storage unit.

SAFETY THAW. The basic principle of safety thaw is to keep food at 40°F (5°C) regardless of the product load. A safety thaw unit has one thermostat that provides refrigeration to lower the temperature to 40°F (5°C) when it rises, plus another thermostat that allows mild heat to raise the temperature to 40°F (5°C) when frozen food makes it fall. This unit will thaw frozen food overnight and still keep it refrigerated after thawing.

These units provide a safe, convenient way to thaw bulk frozen foods for mass feeding. However, due to the increased air circulation, food left uncovered will have a more rapid moisture loss than it would in normal refrigeration.

Food thaws most effectively in containers not more than

2 in. deep. Foil or metal containers are best due to better conductivity. A safety thaw unit costs about 20% more than a medium-temperature refrigerator.

Tables and Accessories, and Carts

Work surfaces, both permanent and mobile, need to be the correct size, height, and construction to improve operating efficiency. Cooks' and bakers' tables can often be procured from an equipment supply company as stock items. Tables that are an integral part of an individually designed unit, such as those connected to a dishwasher or pot-and-pan washing machine or those containing a sink, garbage disposal unit, or other special feature, must be fabricated to fit the specific need. Specifications for fabricated items must include the thickness or gauge of the metal to be used, the method of support, the method of fabrication, and the finish desired.

TABLES AND ACCESSORIES. Worktables are essential in every production area. For greater convenience and flexibility it is better to have some work space fixed and some mobile.

Worktables that are to be in a permanent location are usually 30 in. wide and any length appropriate for the location in which they will be used. Tables more than 6 ft long require added support for durability. This increases the cost of the unit. Long tables also reduce efficiency by adding steps that are required to walk around them. If a space is longer than 9 ft, it is better to purchase two smaller tables or one 6-ft table and a mobile unitized worktable that could be placed where it is needed for the most convenient work arrangement. Allow a minimum of 4 ft of counter space for each worker in the preparation area.

Mobile tables with several shelves and/or slots for 18 × 26 in. bun pans or 12 × 20 in. serving pans provide additional work and storage space for these work centers. Any mobile tables should have good casters with locks on at least two of them.

Table tops should be made of 12-gauge stainless steel; 14-gauge stainless steel may be used if it is properly supported. Stainless steel legs may be attached to the table by means of a fully enclosed gusset or a fully enclosed inverted channel; joints must be seamless for easy cleaning and sanitizing. Tables with tubular legs can be adjusted to the desired height and a shelf can be added, if desired.

Pear- or bullet-shaped feet on tables without casters allow for a small amount of adjustment in table height. Neoprene plugs and gaskets provide a sanitary seal for the neck and bottom of each foot.

If the staff includes a baker and regulations allow it, a wooden-topped baker's table may be used. The wood top is made of 2-in.-thick kiln-dried maple consisting of strips glued and bolted together with steel rods. It should be sanded smooth, sealed, and coated with pure, melted paraffin. However, some sanitarians prefer that bakers use a stainless steel table top and large pastry boards or a composition table top.

Drawers in worktables should be easy to remove and clean. Smooth, rounded edges on drawers help to avoid possible injury. Drawers will operate better if they are on bearings or nylon glides and have safety stops to avoid accidentally pulling the drawer off the runners. The number of drawers depends on the length of the table and the amount of space needed for storage. Several single drawers are more efficient than the same number arranged in three-drawer units because the drawers can be located where they are most needed without bending to reach supplies.

A cabinet for storing condiments, flavorings, and small measuring utensils should be easily available to each worktable. If a table is not located against the wall, the cabinet can be attached to stainless steel or iron rods anchored to the ends of the table, or the cabinet might be anchored to rods suspended from the ceiling directly over the table (Fig. 5.17).

Cutting boards are needed in all work areas, so it is important to provide a spot for storing one or more boards that are large enough for the work to be done but light enough in weight to be taken to the cleanup area to be washed and sanitized properly.

Try to locate a source of water in or near each worktable. It might be less expensive and allow for more flexibility to have a single small sink adjacent to the cook's table rather than to fabricate the sink as part of the table.

Movable bins will be needed for the baker's table and

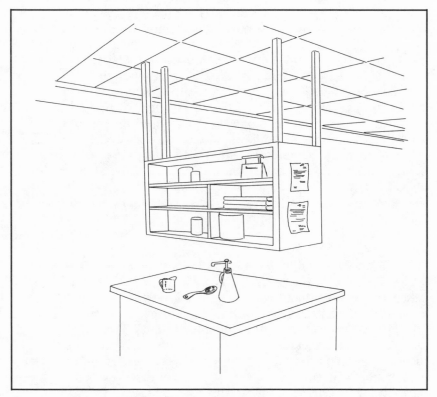

Fig. 5.17. Cabinet suspended from the ceiling.

also might be useful under a section of the cook's table. Specify what size bins are needed to meet the needs of the facility and whether they should be stainless steel or plastic. Plastic bins seem to hold up quite well and are far less expensive than stainless steel.

A safe and easily cleanable knife rack should be attached to the cook's and salad maker's tables. The same kind of rack could be attached to the baker's table to store spatulas and spreaders.

UTILITY CARTS. Every kitchen needs at least two carts to move pans of food from the range to the serving table, to bring food from storage to the preparation area, to move dirty pots from cooking to pot washing, and to carry prepared foods to service. Sometimes a cart is located at right angles to a baker's table to provide extra space for ingredients, or it is used as backup space at the serving line.

The cart that is used to bring canned items from the storeroom may provide the most convenient spot on which to anchor a heavy-duty can opener. A cart used for this purpose should be sturdy and have brakes on two wheels.

Heavy-gauge stainless steel (12–14 gauge) is recommended for the top of utility carts because it is suitable for most loads. Carts to be used for heavy loads should be braced from beneath. There is a cart on the market with a brace at each end rather than supports on four corners. This unit is easily accessible from all areas and has fewer corners to clean.

Good casters will carry the NSF approval. Polyisobutylene wheels are noiseless and will not deteriorate when exposed to steam and water. The hub and swivel should be grease-sealed to protect the bearings against moisture, dirt, and lint. Allow casters large enough so that the cart can move easily, but be careful they do not increase the height of the cart too much for the average employee. Casters 5–6 in. in diameter will enable the cart to be moved easily. These have to be specified because small casters, which make movement difficult, are often standard equipment.

Bumpers are necessary to protect other equipment from being dented. Corner, donut, wraparound, or vertical corner bumpers are available.

6

Space Determination and Layout: Food Serving Area

The layout of the food serving area and the type of equipment chosen have a direct bearing on the productivity of the food service employees who in turn affect operating costs.

The serving area is a work center. It includes the equipment for serving hot and cold foods and beverages and the mobile dish-storage units. There needs to be a tray set-up station in health care facilities and schools and a serving station in restaurants.

Clearly defined traffic aisles 6 ft wide are most important in the serving area, especially when there is a need for mobile dish-storage units. Columns should be eliminated from this area if possible and kept to a minimum in the entire food service department. Locate stationary equipment adjacent to those columns that have to exist.

Health Care Facilities

The serving area in a small (50–100 beds) facility would probably include a tray set-up station, a hot food serving station with tray slide, an angle-ledge upright cart for cold food, and space for the tray delivery carts. A U-shape arrangement that is about 10×10 ft (or 100 ft^2) would probably be adequate. This space would be equal to 1.5–2 ft^2 per patient bed. A facility that is larger than 100 beds would need some type of conveyor unless a large number of the

85

patients ate in the dining room and the tables were set up with flatware and cold foods. The serving line for a facility for 150–350 beds would include a conveyor plus additional serving stations (Fig. 6.1).

The overall width of this serving line could vary from 16 ft to 22 ft. Space is needed for some type of conveyor (1½ ft), both a hot and a cold food serving station (2 ft if parallel, 5 ft if perpendicular), a self-leveling mobile plate unit and maybe a plate underliner unit (2 ft), plus a 4- to 6-ft aisle on both sides of the conveyor. The length will vary according to the number of serving stations needed on either side of the line. Larger facilities may need two hot food serving stations and two cold food stations.

A typical line in a large facility might consist of a tray set-up station (2 ft), work space (3 ft), a cold food station (2 ft), work space (3 ft), milk and ice cream station (2 ft), work space (3 ft), a reject table and work space for a checker (3 ft), plus space for tray delivery carts and the space needed for loading the carts (6 ft). The total space needed on one side of the line would be about 20–24 ft. On the opposite side of the conveyor would be the hot food serving stations, bread, and beverages. This length × width would be in the range of 320–530 ft², or 1–2.5 ft² per patient bed for the tray assembly system. If a cook/chill system was in use and the tray line set-up was used continuously during one 8-hr shift to set up the three meals per patient per day, this same tray line could easily be used in a 600-bed facility; this would amount to 0.8 ft² per patient bed. In addition, approximately 1 ft² per patient bed is needed for cart storage and movement.

If the dishwashing operation is located between the dining area and the serving line and if there is a separation of the dirty and clean areas, some of the clean area could be used for maneuvering and parking tray delivery carts and mobile dish-storage units.

It is important to have almost everything mobile in the serving area so that the system can be adjusted to increased or decreased patient loads, as well as to the kind of equipment that makes it possible to do the most efficient job.

Additional space is needed in the serving area for storing paper supplies and condiments. The coffee station should be near the serving line, but there are disadvantages in putting it *on* the serving line because plumbing and wiring

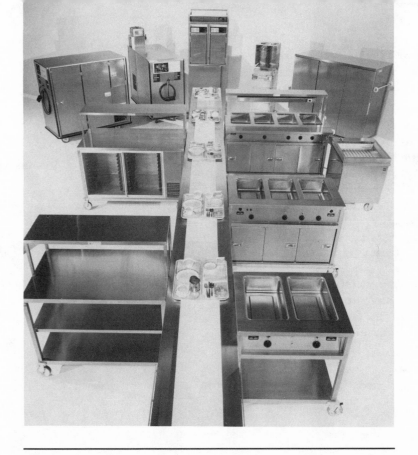

Fig. 6.1. Food serving area. (*Courtesy Crimsco Inc., Mission Hills, Kan.*)

are required for this unit. It is important to keep the serving line flexible.

Schools

School food service programs have changed drastically over the past several years to reflect the eating habits of the clients. Many schools offer several menu choices, often at a number of serving lines to speed traffic flow. Salad bars and other self-serve alternatives are often set up in the dining areas. The decision about the size of all of these units must be based on the kind and number of items that need to be accommodated. Space should be allowed in the dining room for napkins, flatware, and beverages so that the students may help themselves, move at their own speed, and not slow the serving line.

In a self-contained preparation and serving kitchen, about 1 ft² per student will be required. Many schools are using a centralized food service operation and transporting the food to satellite units. The space requirement would then be divided between the two units according to the amount of preparation done in each location. If final preparation as well as serving is done in the remote area, it will be larger and have more equipment than the one in which only serving is done.

Some schools are preplating food and transporting it to the serving site. If this is done, a very small serving area is needed. In fact, this alternative is often used because no serving area is available.

7

Serving Equipment

The information on serving equipment covered in this chapter relates to many types of food service operations. Schools use tray dispensers and may or may not use dish storage units, depending on the extent disposables are used. Restaurants, both fast food and gourmet, provide some work stations with equipment for dispensing flatware, napkins, condiments, beverages, and cold foods. Health care facilities need, in addition to the above, a serving line or conveyor. All types of food service operations use some type of equipment for holding and serving hot food even though most restaurants prepare a number of foods as ordered.

Starter Stations

Starter stations are where nonfood items are assembled on a tray and carried to the dining area, put on the cafeteria tray slide by the customer, or put on a conveyor for assembling and serving food to patients. The starter station that is designed to provide all the tools and supplies needed in the area is one key to efficient service.

Most of these units provide a place for flatware, napkins, racks of glasses and/or cups if used, menus, and condiments. In restaurants, these units are usually a modular cabinet with work space. Health care facilities use a tray-starter cart, which consists of a self-leveling tray-dispensing unit and a frame for shelves as a place to locate condiments, flatware, napkins, menus, and name cards if used. The tray-starter unit should be on casters and should move easily (Fig. 7.1).

Fig. 7.1. Tray-starter cart. (*Courtesy Crimsco Inc., Mission Hills, Kan.*)

SELF-LEVELING TRAY DISPENSERS. The tray dispenser is a mobile cabinet sized to fit the trays being used. It should contain a self-leveling mechanism that can be adjusted at the field site. Nonmarring, swivel ball bearing casters with foot-operated brakes should be specified. Tubes for saucers and bread plates may or may not be a part of this unit. A dispenser for several cylinders of flatware may be attached to the top of this unit, especially for use in a cafeteria setting.

Chilled Serving Equipment

Chilled or cold food should be held and served from chilled equipment to prevent spoilage and maintain quality.

REFRIGERATED CARTS. A refrigerated cart with a working height countertop and space for two or three 18 × 26 in. pans is useful for setting up trays. The refrigerated space below can house 12–18 pans of ready-to-serve foods. The cubic feet of refrigerated space can be counted as part of the total refrigeration. Since this unit meets two needs, it is not as expensive as it seems. Many carts have two shelves above the countertop for pans of cold foods used during the serving period.

Keep a refrigerated unit near 40°F (5°C) and be sure the cold air is distributed evenly and the humidity is adequate to keep the products crisp and attractive. There should be a thermostat control and a thermometer to indicate the inside temperature; both should be easily accessible. A ¼- to ⅓-hp compressor with 120-volt current is adequate.

The stainless steel top of this unit must be firm (16–18 gauge), with the top and sides well braced. The sides and the ends may be covered with 18- to 20-gauge stainless steel.

CLOSED CABINETS WITH COLD PLATES. A closed cabinet without refrigeration can be cooled by the use of cold plates. Backup refrigeration located nearby is an important factor in deciding on closed cabinets with cold plates. Such cabinets are much less expensive than refrigerated carts, which literally take the place of an equal amount of backup refrigeration.

The cold plates are constructed of stainless steel and contain a sealed-in chemical refrigerant. They are about 13 × 18 in. and weigh approximately 22 lb. The cold plates have to be put in the freezer overnight to freeze the refrigerant. Two cold plates can cool a closed cabinet approximately 6 ft high. An additional cold plate is added protection in an area that might be subject to delay in serving.

A typical cabinet can hold as many as six 18 × 26 in. pans with a 5-in. space between each pan. For cold foods to be located within easy reach of the station worker, the bottom pan should be about 30 in. above the floor and the top pan about 60 in. above the floor.

OPEN-RACK CARTS. Rack carts that are equipped with universal-angle tray slides can be used for 18 × 26 in. pans of cold food. Tray slides may be welded to the frame, or they may be removable and adjustable to 2½-in. modules. Each wide tray slide is designed to accommodate one 18 × 26 in. pan, two 14 × 18 in. pans, two 12 × 20 in. pans, or one 20 × 20 in. glass rack. Removable tray slides can be cleaned in a dishwasher but may be a hazard if they become loose and slip out too easily.

Carts with tray slides are available in aluminum or stainless steel. Stainless steel models are about twice as expensive as aluminum models and weigh about three times as much. Either type of cart can do the job, but a stainless steel unit will take more abuse with less damage.

The usual size for a cart of this type is 65 × 29 × 20 in. A cart that is open on the 29-in. side is preferable on a tray line so that the reach to the far edge of the sheet pan is not longer than an arm's length (Fig. 7.2). If the cart is open on the 20-in. side, the pans must be turned during the serving period in order to reach the food on the far end of the pan. This type of cart provides space for some of all types of cold food needed at one time. Extra supplies of the same foods can be available on food file–type shelving or a similar type cart in a refrigerator nearby.

Fig. 7.2. Open-rack tray cart. (*Courtesy Lakeside Mfg. Co., Milwaukee, Wis.*)

Specify the width of the opening and angle ledges, the height, the material (stainless steel or aluminum), donut-shaped or perimeter bumpers, tubular pan stops, the size of casters, and brakes if desired.

MOBILE CARTS WITH ICED WELLS. Refrigerated or iced pans or wells are useful for serving large bowls of salads and salad dressings or for containers of many salad items for a self-service salad bar. Rounded corners plus seamless construction minimize sanitation problems. Seams should be Heliarc welded (a combination of gas and electric welding). Recessed handles permit mobile units to be placed flush against the wall or other equipment. There should be access to a drain for removing the water from melting ice. A source of ice such as an ice machine should be available to provide refills.

FROST-TOP UNITS. Self-contained refrigerated frost-top units for salads and desserts are available in various lengths (Fig. 7.3); they can be built in or mobile. A refrigerated storage unit is usually below for backup storage of cold food items. This type of unit is appropriate for a cafeteria line, as it provides space for displaying and serving cold food items.

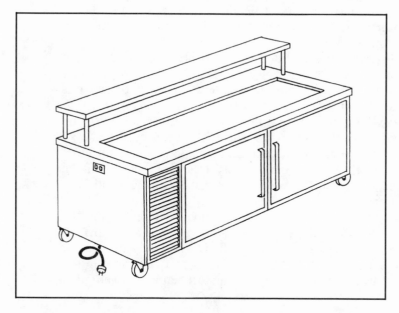

Fig. 7.3. Frost-top unit for salads and desserts. (*Courtesy Caddy Corp. of America, Pitman, N.J.*)

Be aware of the gauge of the stainless steel, the thickness of insulation, the width and swing of storage-space doors, and if the unit is self defrosting. Specify the length and width of the unit desired and the electrical and plumbing requirements.

Heated Serving Equipment

Hot foods should be held and served from heated equipment. The size or number of pan openings on hot food serving equipment should be determined by the varieties and kinds of foods to be served. Mobile units are available in 2- to 4-well units. Fabricated units contain as many wells as are specified.

Most units have the wells located side by side, but there are some units with two or three wells located end to end. These units are useful on a serving station or tray line where there are more foods than can be put into a 4-well serving table. In a situation like this, the 4-well hot food serving station can be located perpendicular to the line, and the narrow unit with wells end to end can be parallel to the line making it fairly easy to serve from either unit.

Hot food serving units vary in height from 34 in. to 42 in., but the most common heights are 35⅜ in. and 37 in. Variation in height is a result of the size of the casters on mobile units.

Different parts of hot food units are made of different gauge stainless steel; a heavy gauge is used for the top, a light gauge for the side. Most units have other parts or sections that are reinforced for added strength. There also are variations of the gauge of stainless steel in the same parts by different manufacturers; the range is from 14 gauge to 22 gauge. The top of the unit should be made from one piece of stainless steel with no seams or crevices. The corners of the hot food wells should be covered so that no dirt can collect in these areas.

Strip-type heating elements provide even heat under the whole well. Radiant heating elements provide heat through the bottom and walls of the well. Other kinds of heating elements are a U-shaped rod and an exposed element in the shape of a figure eight. A unit that surrounds the well gives more constant heat than an open unit under the well. Heating elements that are enclosed are less likely to cause scorching of food. It is important to have a drain if water is needed to create moist heat under the hot food.

The wattage for heating elements ranges from two 150-watt units in one element to 1,000 watts per element. Separate controls for each well minimize the amount of current needed at any one time, eliminate unnecessary heat in the surrounding area, and keep each type of food at the proper temperature at the time it is being served.

The thickness of insulation in hot food units varies from ½ in. to 2½ in. All units that include heated sections for storing extra pans of foods below the serving area are insulated, although the doors to these areas may or may not be insulated.

Some doors are attached by piano hinges, which let a door fold back out of the way. Other doors are attached by pin hinges, which can be taken apart so that the door can be removed for cleaning the inside of the cart. Some doors have self-sealing magnetic gaskets.

Casters are 4–8 in. in diameter. Most are ball bearing swivel casters, two of which are fitted with locking brakes. For mobile heated carts, 5- or 6-in. casters are most appropriate; 4-in. casters are too small, 8-in. casters are too large.

Mobile units that have recessed handles can be located flush against a wall or other equipment. These units must have bumpers around the perimeter or on the corners to protect the cart and equipment that might be hit by another cart. Accessories include serving pans, front and/or back-drop shelves in stainless steel or wood, tray slides, over-shelves, sneeze guards, infrared lights, additional bumpers, reel cord assembly units, push handles, and brakes. One company makes a hot food serving cart with an overhead shelf, air shields, plus infrared lights to keep the food hot while being served.

Specify the number of serving wells according to the number of hot items that will be served, the number of thermostats needed, the width of tray slides if needed, the size of casters, the voltage and phase, and any accessories desired. Be aware of the thickness of the insulation and the gauge of stainless steel on all parts of this unit.

HEATED DISH-STORAGE UNITS. Hot dishes are the secret to serving hot food that has just been cooked, not cooked and chilled. There are heated self-leveling units for storing vegetable dishes, soup bowls, plates, and underplates. Thermostatically controlled units that are adjustable from 165°F (75°C) to 200°F (95°C) provide a means of having the dishes at the temperature suitable to your facility. The switch and thermostat should be easily accessible.

If the unit is enclosed a door on one side is valuable so that the inside is accessible for cleaning. Springs should be accessible and easy to adjust as needed. Wraparound or donut bumpers provide protection for the unit and equipment that might be hit by this unit.

Strong dish guideposts are needed in plate storage units to prevent breakage, since several plates usually stand above the top of the tube. These guideposts may be made of stainless steel or stainless steel covered with vinyl to prevent marring and scratching the plates. The diameter of a plate, the height of a dozen plates, as well as the weight of a plate is needed in order for the company to set the spring correctly. Each tube holds approximately 90 plate covers, 55 underplates, or 72 plates: more, if the newer, thinner china or some of the other thin, nonporous, hard-finish tableware is used.

There is a cabinet-type unit with rods as dividers instead of tubes. The rods can be firmly located to provide space for various sizes and shapes of dishes; even more than one size or shape can be accommodated in one unit. Covers of various types are available for most mobile dish-storage units.

When determining the type of unit to be used, be aware of the gauge of the stainless steel tubes or wraparound stainless steel walls and the adjustability of the springs. Specify the numbers of tubes needed (available in groups of 2, 3, or 4 tubes per unit) or the size of the cabinet with rods for separators, the voltage and phase, push handle if desired, the size of casters, brakes on two wheels if needed, wraparound bumpers, and a separate switch for each tube if the units with tubes are purchased. It is less expensive to purchase units with several tubes than it is to purchase several units with one tube each. A dispenser that is enclosed with stainless steel around the tubes will cost much more than the tubes with circular heating elements (Fig. 7.4).

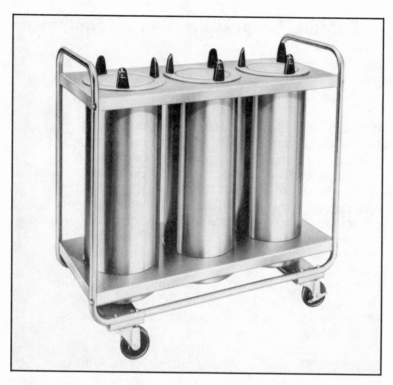

Fig. 7.4. Three-tube dish storage unit. (*Courtesy Lakeside Mfg. Co., Milwaukee, Wis.*)

CONVECTION-HEATED DISH-STORAGE UNITS. These mobile self-leveling dispensers provide forced convection heat to warm underplates or the heavy plates that are used in some restaurants. The forced convection heat reduces heating time as the heated air circulates evenly throughout a well-insulated cabinet. Underplates or heavy plates can be heated to 225°F (145°C) in 1½ hr and held at that temperature until dispensed. The units are thermostatically controlled to prevent overheating. These units are insulated on all sides and have an insulated cover.

Room-Temperature Serving Equipment

Many items are served at room temperature. Efficiency of handling is the most important consideration in selecting room-temperature units.

SALAD AND DESSERT PLATE STORAGE UNITS. Mobile storage units for salad and dessert plates need not be heated or self-leveling because the purpose is to get the clean dishes from the dishwashing area to the work station where salads and desserts are portioned for serving. The factors that should be studied when selecting this equipment are ease of movement, the correct amount of storage space for the number and size of dishes to be used in each area, and a means of covering the unit during periods of storage (especially when the floor is being cleaned).

These units are 23–36 × 10–24 in. Some units contain heavy, plastic-coated wire dividers that can be easily moved to provide the space needed to separate stacks of dishes. However, two or three dividers will require 2–4 in. of space inside the cart. A cart 36 in. high will have space for a stack of approximately 72 salad or dessert plates, or 62 fruit or sauce dishes. If 144 7-in. dishes and 144 5-in. dishes are needed in the area for portioning salads and desserts, a cart with 24 in. of inside usable space, or with space available on two sides of a partition, or two carts that have 14 in. of inside usable space will be needed. Carefully examine the cover or closure for carts of this type. Some covers on the double carts are located so that they are in the way when the worker is trying to use dishes from both sides of the cart. There is a well-designed cart on the market that has a top cover that slides into a vertical position behind the dish shelf and a front cover that slides horizontally under the dish shelf when not in use. A push handle at one end of the cart is a must if a cart is to be moved from one area to another. Carts of this type should never be so big or heavy that they cannot be moved easily when fully loaded.

CUP AND GLASS STORAGE UNITS. The most efficient way to store cups and glasses is in the plastic racks in which they are washed. These racks can then be placed in an open cart that has a cover; a closed, self-leveling mobile dish-storage unit; a closed unit with the door opening so that empty racks can be placed on the bottom shelf as the rack is emptied; or an open unit with a solid shelf and space below for storing the empty racks (Fig. 7.5).

Fig. 7.5. Open cup rack.

Conveyors

In the context of this chapter, conveyors (gravity-feed or motorized) are discussed in relation to serving food on trays; however, some of the information applies to vertical conveyors or dumbwaiters that are needed to move food from the production area to the serving area or to move soiled dishes from the dining area to the dishwash area. Factors that should be considered when designing a system are speed, size of tray, length of conveyor needed, electric current available, ease of cleaning and maintenance, and, of course, cost.

GRAVITY-FEED CONVEYORS. As the name implies, gravity is the force that moves objects on the conveyor skate wheels or rollers. Although no external current is required, outlets may be needed for hot food serving tables; heated dish, bowl, and underplate units; and refrigerated food service and storage units. The height and slope of this type of conveyor are usually adjustable to provide some degree of speed control; however, employees have to move the trays along as they complete their part of the process.

The conveyor units are available in 5-, 10-, 15-, and 20-ft lengths; long conveyors need leg support every 5 ft. The table should be made of 14-gauge stainless steel and the space between the sides should be about 1 in. wider than the trays conveyed. The support legs can be stainless steel or aluminum and have bullet feet.

Fig. 7.6. Skate-wheel conveyor unit. (*Courtesy Caddy Corp. of America, Pitman, N.J.*)

Skate-wheel units consist of two parallel rows of wheels that move the trays the length of the conveyor (Fig. 7.6). This type of conveyor is useful for moving trays of soiled dishes in the dishwashing area, for a built-in trough can be recessed into the unit between the rows of wheels. Food from the trays can be emptied into the trough and then flushed toward a disposal unit.

Roller conveyors have rollers that extend the full width of the conveyor bed or table and offer more stability for trays (Fig. 7.7). These units cost about 10% more than the skate-wheel units, but both are less expensive than motorized units.

Fig. 7.7. Roller conveyor unit. (*Courtesy Caddy Corp. of America, Pitman, N.J.*)

MOTORIZED CONVEYORS. If motorized units are used, the number and location of outlets as well as the voltage specified for each is very important. The Hubbell and/or Arrow Hart number should be provided for the engineer who designs the electrical system for the conveyor line. The control panel may include electronic speed control as well as overload and underload voltage protection. The length, height, and direction of movement will determine the placement of the start/stop switches at both ends of the unit.

Mechanically driven conveyors are continuous belts, the fabric for which should be specified depending on the intended use. A slat conveyor has a plastic slat belt mounted on a 14-gauge stainless steel table (Fig. 7.8). It is said to require less maintenance than fabric belts. This unit can be designed to turn corners easily and is suggested for a continuous washing system.

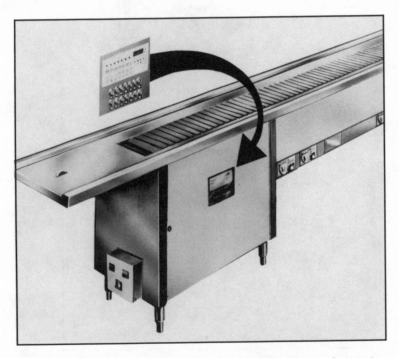

Fig. 7.8. Slat conveyor unit. (*Courtesy Caddy Corp. of America, Pitman, N.J.*)

Fabric belts should be made of sturdy material that is relatively nonabsorbent to grease and liquids, should be nontoxic, and should have no raw edges. The belt support and drive mechanism should be easily cleanable; therefore a wash system with proper drains and motors should be specified. Cleanout drawers are needed at both ends of a long conveyor, and if panels are designed to cover the working parts of the unit, they should be removable for cleaning purposes.

One of the newest types of motorized systems has a set of conveyor bands as part of the unit that has built-in space for a frost-top unit with refrigerated space below and space for a configuration of 12 × 20 in. pans (Fig. 7.9).

Another new concept is the use of a chain link drive with removable tray carriers. The width of both these systems ranges from 4 ft to 5 ft. The length can vary from 12 ft to 30 ft depending on the size of the operation and the type of menu. These units can be prewired and preplumbed for one-point service connections. An overshelf is available for loca-

Fig. 7.9. Frost-top Coldveyor system. (*Courtesy United Service Equipment Co., Murphreesboro, Tenn.*)

tion of nonfood items. These conveyors can be used with the cook/chill system or in some cases be used with the conventional hot and cold carts that are set parallel with the conveyor base. The base can be on casters or a stationary pedestal.

Tray Delivery Systems

A 1970 Hill-Burton survey evaluated problems of food distribution systems. The survey found that if passenger elevators are used to move food carts, the carts should be enclosed to reduce contamination. Also, keeping food at the right temperature while transporting it is a major concern.

There are several systems for transporting food in either heated or chilled carts. A new concept being used in some very large facilities is the cook/chill production system interfaced with blast chilling or freezing, transporting in cold carts, and then rethermalizing. The cooked food is quickly chilled to 36°F (2°C) in a blast chiller and stored in serving pans at 38°F (3°C) for 4–5 days or pumped directly into plastic bags and sealed; sealed bags of chilled food may be stored up to 45 days at 36–38°F (2–3°C). In either case the food is plated in a chilled state from a tray line of refrigerated units. The plated food is held under refrigeration until it is to be served; it is then delivered to a galley where the hot food is rethermalized. There are several rethermalization systems available to choose from.

REFRIGERATED CART WITH SEPARATE CONVECTION OVEN RETHERMALIZATION. Each polyurethane-insulated transport module (cart) is chilled by contact with a compressor in the production area or by placing the cart in a walk-in refrigerator; the compressor is not transported (Fig. 7.10). The interior of the cart is cold when loaded with trays of food that are plated from chilled stations on the tray line. The delivery cart can continue to be chilled in the galley by being rolled into contact with a compressor designed for that purpose; the cart is lightweight and easy to move. The food is served on standard china plates and covered with closed, unvented covers to retain moisture. Each cart holds 20 trays, which are set on 4-in. centers.

Fig. 7.10. Refrigerated transport module. (*Courtesy United Service Equipment Co., Murphreesboro, Tenn.*)

Direct physical relationship between the tray delivery cart and the rethermalizing convection oven is achieved through specially designed door-mounted oven racks. This ensures the efficient reheating of meals and error free assembly of trays.

The double-unit oven will accommodate 5 serving pieces per rack on each of 5 racks in combinations of entrees, soups, cereals, beverage pots, and 12 × 20 × 2½ in. hot food serving pans for a total of 25 pieces per oven unit (Fig. 7.11A, B). Push buttons marked A.M. and P.M. allow for a specific preset time for rethermalization in the oven. A beeper sounds when the cycle of heating is completed.

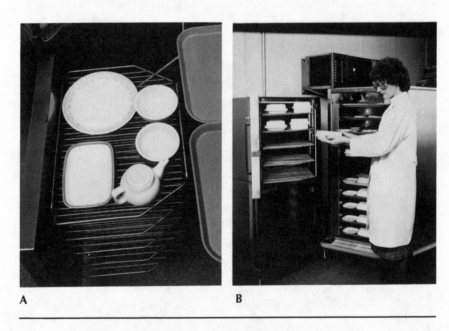

A   B

Fig. 7.11. Refrigerated transport module and convection oven. **A.** Loading pattern for oven racks; **B.** Transferring chilled plates to oven for rethermalization. (*Courtesy United Service Equipment Co., Murphreesboro, Tenn.*)

REFRIGERATED CART WITH ATTACHED CONVECTION OVEN RETHERMALIZATION. This is a highly mobile system for hot and cold foods and hot beverages. Each cart is 32 × 58 in. and holds 20 trays. The convection oven permits fast efficient heating of chilled entrees and soups from 40°F (4°C) to 170°F (75°C). The refrigerator unit keeps cold food cold. A hot beverage heater and a small freezer (holds up to 23 ice cream cups) are available if specified. The hot food has to be added to the tray of cold food in the serving area.

RETHERMALIZATION BY RADIANT HEAT. If this system is being used, the food is served from chilled stations onto specially designed porcelain plates and covered with stainless steel lids. The plates are then placed on a wire rack that is attached to the end of each lightweight, enclosed cart. A heavy, hard plastic form is placed over the wire rack of plates. The wire rack is then rolled into a cavity that produces radiant heat. The food is heated in 17 min. The soup is heated in a microwave oven and the beverage is served from the beverage dispenser or heated in the microwave. The carts that contain the trays with salads and desserts may be held in a walk-in refrigerator or taken directly to the serving area. The ice cream and milk are served from the freezer and refrigerator in the area.

RETHERMALIZATION INSIDE A REFRIGERATOR. This system consists of specially designed servers and compartmentalized insulated covers; rethermalization carts that will hold 16, 20, or 24 servers; and special refrigerators in which carts are held below 40°F (4°C). The food items to be heated are placed in specific cavities on the server. Each server has three cavities for hot food; spaces for three cold items; and a wing for menus, flatware, and beverages. The use of a full cover for food eliminates the need to wrap cold items. A single button activates the connection between cart and refrigerator; conduction heating components become very hot and heat hot foods while the refrigerator component in the server remains cold keeping cold foods cold. No temperature rise occurs within the refrigerator during the heat cycle. A timer sounds and flashes when the food reaches the correct temperature; this takes 35 min. The carts are removed from the refrigerator in the galley, the trays are completed and delivered to patient.

The system can also accommodate complete cold meals, which are placed on the shelves backwards to keep the activating switch from being tripped. Specially designed serving dishes or disposable dishware can be used with this system. A contract may be made with the company for provision of disposables, repair of equipment, and replacement of trays.

THERMAL AND COLD PELLET SYSTEM. This system uses thermal and cold pellets for the hot and cold foods. A convection-heated, self-leveling unit is needed for heating plates and/or plate holders. This system allows for flexibility of the number of trays needed in any one area and provides some flexibility in adding a few trays at a time as needed.

The pellet is 18 × 7 in. and fits into a standard 20 × 15 in. tray. Hot food is kept hot by being served on hot plates that are heated in the convection-heated, self-leveling unit and set in a thermal pellet. Hot soups and beverages are served in insulated bowls, mugs, and/or cups. Cold food is placed in a cold pellet that contains a sealed-in gel that has been chilled about 1 hr in a dispensing freezer. The cold foods can be assembled and covered with a clear plastic cover, placed on 26 × 18 in. trays, and put into the refrigerator until serving time.

The tray assembly procedure can be handled in the usual manner: set the cold pellet and thermal pellet on the tray and then add the hot soup or beverage, if called for. All food on the tray is then covered. Inexpensive carts can be used for delivery of the trays.

A special rack is needed for air drying the plate covers after being washed.

COMPOSITION DOME COVER AND BASE SYSTEM. With this system a foam-insulated plate base and cover captures and maintains plate and food heat. Insulated bowls and mugs are needed for soup, hot cereal, and hot beverages. Plates and the food on the plates must be hot when put into the base because there is no other source of heat. Speed in service is necessary for satisfactory results. This is a relatively inexpensive service, a little over one-half as much as for the hot-and-cold cart or thermal disk system.

Racks for holding the bottoms and covers are necessary

for the system to work in a large facility. Since the spaces between the divisions on the storage rack must be adequate for the bottoms and covers that have been chosen, these items must be purchased as a complete system.

There are many other tray delivery systems that have been in existence for a number of years and are still available. Some of these systems meet the needs of some facilities better than the rethermalization system.

THERMAL DISK SYSTEM. With this system, an enclosed metal disk is heated in an oven or self-leveling heated mobile unit. The plate of food is then set on the hot disk and a plastic or stainless steel cover placed over it. However, hot soups and beverages, as well as frozen desserts, need to be served in individual insulated containers. This thermal disk system has advantages over the hot-and-cold cart in that it can be started with a few units. The size of the serving cart can be determined by the number of people to be served in a specific area. If you need to serve only a small number of people, thermal disks may be warmed in a regular oven, but space should be allocated for the addition of a special oven or self-leveling dispenser heater to be added when needed.

An enclosed unheated cart is a sanitary precaution in transporting trays if the salads and bread are not covered. If you are considering the thermal disk system, the tray cart should have clearance between shelves for the beverage containers. The thermal disk system is nearly as expensive per tray as the hot-and-cold cart when calculated on the basis of 20 trays to a cart. This includes the thermal disk cover, insulated beverage servers, and other special items. However, when the thermal disk system is chosen, there is the advantage of being able to add single units as needed. There are several different thermal disk systems available. The disks range in weight from 8 oz to 24 oz. Consider the total weight of a tray plus disk, food, and dishes when selecting a system.

THERMAL TRAY SYSTEM. This system consists of three main elements: thermal plastic serving tray with various size compartments, disposable plastic food holders that fit into the compartments, and specially designed carts that weigh much less than hot-and-cold carts.

The serving trays are designed to be vertically interlocking, and when they have been filled, they are stacked one on top of the other; the top tray is always covered with a lightweight plastic lid. Since similar food items are aligned, the stacking process forms "thermal columns" that maintain temperatures, either hot or cold. The manufacturer provides the equipment, the procedure for operation and equipment maintenance, and a controlled inventory of disposable food holders. The facility is charged monthly on the basis of disposable inserts needed for the number of meals served, so there is no initial cost.

ELECTRICALLY HEATED TRAY SYSTEM. Another type of cart available has a rechargable 12-volt lead battery as the heating source. Heating elements are built into each tray with contact points that connect to complementary heating terminals within the cart. A drive system increases and decreases drive power and output up and down inclines. An electronic sensing device built into the front of the unit will tell the cart to stop when necessary. Each of the tray slots and trays has an assigned number on a logic panel. By pushing a button for a particular tray the plate and/or bowl on that tray can be heated. All trays can be programmed to heat prepared items for the individual and keep the hot items hot until served. Since the heating element heats only a specific area on the tray, cold items will not be affected. The trays are used with a specific type of disposable dishware.

INTEGRAL HEATING SYSTEM. With an integral heating system, foods are prepared, frozen, and stored in a central kitchen. The preplated frozen meals are then placed in special dishes that consist of a plastic outer shell and a porcelain-coated steel inner compartment. The dishes are placed on trays and the trays are placed on rails in the integral heat unit. Each unit holds 24 meals.

Electrical energy from push-button controls comes through rails to electrodes—small metal buttons—on the bottom of each plastic outer shell. The energy passes through the buttons to a thin-film resister on the bottom of the inner dish, where it is converted to heat and transmitted through the dish to the food. The food is then served in the same dish in which it was frozen and heated.

The system offers high heating efficiency—90% of the

generated heat is delivered to the food. A 10-oz frozen meal can be made piping hot in 18 min, while the interior of the integral heat unit and the thermal shell of the dish remain cool to the touch.

ENCLOSED UNHEATED CART SYSTEM. Carts that can be used on regular or special elevators, dumbwaiters, or tray conveyors are needed to deliver the trays with thermal disks or the composition dome cover and base system to the areas where patients are located. An enclosed unheated cart with space for 8-20 trays is frequently used when trays are delivered on the regular elevator or to patients located on the same level as the food service operations.

One cart has composition doors that are lightweight and very durable. They have a spring opening that controls the position of the door when open and an airline latch that is more durable because there are fewer moving parts. Another cart is designed as a dual cabinet with a loose pin-hinge arrangement that permits the cabinet to be loaded on both sides, closed, and transported to the eating area; the halves then can be moved to different serving areas. There is no need for doors on this type of unit. The tray supports are removable and therefore the unit is easy to clean.

Some large facilities have an enclosed dumbwaiter designed to handle carts that carry eight trays at one time. This system usually provides an easy method of putting the cart on the elevator and automatic ejection of the cart at the proper floor level.

HOT-AND-COLD CART SYSTEM. A system preferred by some operators is that of the cart with divided trays that are completely assembled in the kitchen and slid into the cart. Rubber or composition baffles separate hot from cold on each tray. These baffles do become worn and may have to be replaced. This is expensive but could be done by a maintenance man.

Another system merely separates hot foods from cold foods, with cold foods on a tray in the cold side and hot foods in a drawer or on a small tray in the hot side. A disadvantage of this system is chilling of the silver, and possibly of the cup, if it is placed on the cold tray. A greater difficulty lies in the fact that hot and cold must be matched after the cart reaches the patient floor. This is of more concern in hospital than in other health care facilities, for in other health care facilities the patients usually stay long enough for the staff to become familiar with the diet prescriptions. It is usually necessary for a dietary employee to go with the cart to see that foods are correctly paired.

DRY ICE–REFRIGERATED TRAY CART SYSTEM. This is an enclosed stainless steel cart with a section for dry ice, so no electrical outlets are necessary. The cart can be used for moving a completed tray of food to an area where there is a microwave oven for heating hot food.

REFRIGERATOR-MICROWAVE TRAY CART SYSTEM. This cart contains two refrigerated cabinets with foamed-in-place polyurethane insulation. These cabinets hold 20 trays. In the area between the two refrigerated cabinets there is space for a microwave oven and two pullout drawers that can be used for storing cups and/or beverage servers. This cart will keep the cold food cold until the patient is ready to be served. The microwave oven will heat the hot food quickly when it is needed whether at the regular serving time or for delayed service. The tray slides can be removed without tools and placed in the dishwasher for sanitizing.

CARTS ATTACHED TO AUTOMATIC CONVEYORS. This system is designed to move supplies vertically and horizontally throughout a facility. Its off-the-floor transportation may reduce cross contamination. The automatic cart transportation system improves the movement of food to patients and the movement of soiled dishes and trays to the dishwashing or sterilizing area. The automatic vertical conveyor system is used for moving all supplies throughout the facility.

In the dietary department, clean, empty carts are automatically delivered and stored on the monorail. At mealtimes the carts are automatically brought to a position adjacent to the tray makeup areas; the trays are set up, covered, and placed on the shelves of the cart. The loaded unit is then sent to the specific destination within the health care facility complex.

All carts should have push handles, corner or wraparound bumpers, nonmarring neoprene casters, and doors designed for easy access to the inside of the cart for cleaning.

Provide a drain in the bottom so that the cart can be sanitized and drained. Some companies provide optional brightly colored special trim for pediatric tray carts.

Keeping food at the right temperatures and preventing contamination of the food while transporting it throughout a facility is of major concern in a dietary department. Important factors to consider when choosing a system are the weight of the transporting cart, the number of motions necessary to completely set up and serve the food, the chances for mismatching food required by diet order, the length of time for rethermalization versus the problems of keeping cold food cold, the amount of labor required to accomplish the job, and the cost of the total delivery system.

Choose a system and specify every item as a part of the whole unit to provide efficient service: trays must be large enough for the dishes you plan to use; the space between the tray slides must be wide enough for a tray with a carton of milk, a beverage pot, and a glass to slide in easily.

Bulk Food Transportation

Satellite systems in school food service, mobile meals for the home-based, and party service in any operation all make bulk food transportation a necessity.

PAN CARRIER. Thermal portable carriers are usually built of 18-gauge stainless steel, are insulated with polyurethane foam, and come in several varieties:

1. With the top solid or recessed to accommodate serving pans.
2. Heated or unheated.
3. With or without cold keeper.
4. With or without carrying dolly.

The carriers, heated or chilled, may be used for transporting food to different areas or buildings. The cabinets hold up to four 12 × 20 × 2½ in. steam table pans or six 18 × 26 in. pans.

The heated portable units have a thermostat and may be plugged into any 110/120-volt single-phase 60-cycle outlet.

(It is important to have a space to roll up and store the cord on the cart.) The temperature loss in 10 hr is only about 18°F (−8°C) even without intermediate heating.

Thermal portable carriers have carrying handles on both sides that will fit into a pocket or groove and can be removed or folded out of the way. A base with 5-in. casters makes it possible for this unit to become a food transporter.

Other bulk food containers are made of molded plastic that is insulated with foam to retain hot or cold temperatures for up to 5 hr. These units stack securely for easy transport. They are designed to accommodate full-size (12 × 20 in.), half-size (12 × 10 in.), and even smaller serving pans. A belt buckle assembly is available for carrying three components of this type of carrier.

MEAL CARRIER. There is a meal carrier that will hold 12 plates. It has a tightly sealed cover, full-length hinge on the door, a wide-opening door, foamed-in-place insulation, and is made of aluminum without sharp corners. It retains the temperature of the food but does not retain odors.

MEAL PACKS FOR MOBILE MEAL SERVICE. These are insulated metal containers that can be individually heated electrically, or Pyrex or aluminum foil inserts that can be heated in an oven. Disposable dinnerware also can be used instead of Pyrex or foil. If Pyrex inserts are used, a duplicate set will be needed for every person so that the unit may be left with the individual overnight.

INSULATED PLASTIC TRAYS FOR MOBILE MEAL SERVICE. Lightweight units with good heat retention are insulated plastic trays with divided space for entree, dessert, salad, and bread. Paper or styrofoam inserts are placed inside each section of the tray. A duplicate set is needed for every person so that the tray may be left with the individual overnight.

PREPACKED SYSTEM FOR SATELLITE SERVICE. With today's new types of equipment, new serving methods, and new foods, food can be prepared ahead in a central kitchen or procured from an outside supplier. When you need the food for serving, the precooked meals, individually and tastefully presented for serving, are quickly and easily reheated. The food is heated by diffused infrared rays that do not penetrate the food but heat the stainless steel cover, which in turn heats the food.

A packaging conveyor is needed if a large quantity of meals are to be sent to distribution centers. Plastic and foil containers are used for packaging and transporting cold food that will be heated in a reconstituting oven. Dollies and baskets are useful for transporting the hundreds of lunches packed this way.

Coffee Makers

Primary considerations when selecting a coffee maker are the amount of coffee needed, whether a lot of coffee is needed in a short period of time or a small amount of fresh coffee is needed regularly over a long period of time, and the space available for installation.

Coffee urns are made to brew large amounts of coffee by using the drip method of preparation. The units usually consist of a hot water storage tank that surrounds two coffee storage containers. There is a wide variety of sizes ranging from 3 gal to 20 gal.

Coffee urns may be manual, semiautomatic, or fully automatic. Manual models require that the water that is boiled in the outer jacket be drawn off and poured over the grounds. Semiautomatic models siphon water over the coffee grounds through a spray nozzle, automatically controlling the amount of water. Automatic models are equipped with a switch for push-button brewing, a timer to control the brewing cycle, and a flow-control valve for water intake. Even coffee agitation is automatic. In most cases there are two urns connected to a hot water boiler. Two batches of coffee can be made simultaneously, or staggered, or only one urn can be used if the needs are less than usual. Automatic brewers correctly control the contact time of water with the coffee grounds, water temperature, and water volume. They complete the cycle in 4–6 min.

New urns (twin or single) can brew as little as ½ gal or more as needed (Fig. 7.12). The newer urns have a selector switch, preset cycle, agitation system, coil to heat incoming cold water, automatic jacket refill, etc.

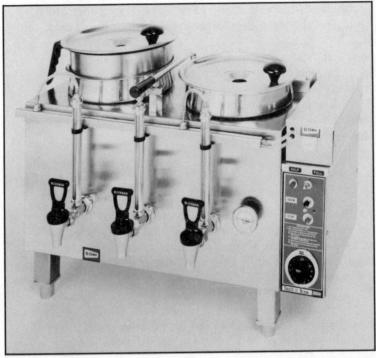

Fig. 7.12. Micro-Brew twin coffee urn with optional hot water. (*Courtesy Blickman Equipment Corp., Union City, N.J.*)

The dispenser system for freeze-dried coffee does not use filters and there is no disposal of grounds. The flavor is presumed to be consistent. Hot tea and hot cocoa can be available from the same machine. The machine has to be connected to a hot water line and an electrical outlet. The equipment is available from some companies, without charge, if a contract to purchase coffee and tea from that company is agreed upon.

The dispenser system for frozen liquid coffee can produce a cup of coffee in 2 sec. The flavor of every cup is said to be identical. The dispenser will hold enough coffee to make 400 cups, yet there is never any left over coffee and the coffee is never stale or bitter. The machine is usually available without charge provided there is a contract for purchasing the frozen coffee from the company that provides the machine. A water connection is necessary. The equipment is easy to clean.

Toasters

Heavy-duty rotary or pop-up or pop-down four-slice toasters are the choices available for preparing large amounts of toast. Two heavy-duty pop-up or pop-down four-slice toasters may provide more flexibility than one rotary toaster. During busy periods, either type of toaster could be loaded with eight slices of bread at one time. However, if only one or two slices are needed, the pop-up or pop-down toaster is faster and requires less electrical current. Two four-slice heavy-duty electrical toasters cost almost as much as an eight-slice electric rotary toaster.

Some toasters can be used for toasting buns and bagels as well as bread. A special switch can be turned to bun-toasting position, which turns off about half the rear heating element. This reduces the amount of heat applied to the crust surfaces of the bun, which allows them to warm but not burn (Fig. 7.13).

Fig. 7.13. Heavy-duty rotary toasters. (*Courtesy Hatco Corp., Milwaukee, Wis.*)

A rotary toaster can have front or rear delivery. Toasters should have removable crumb trays for cleaning.

It is useful to have an automatic voltage-compensating feature as part of an electrical timing system to automatically maintain the toast color setting regardless of voltage fluctuations. Another desirable feature is a timer with an automatic temperature-compensating device to maintain desired toast color through any number of successive toasting cycles. The heating units should be of the open ribbon resistance alloy type. It is important to select a toaster that provides easy access to component parts when service is needed.

Toasters work more efficiently on 208 or 220 volts than on 120 volts. There are 8–10 models of rotary toasters to choose from. Specify bread or bun or a combination of both. Specify speed or the number of slices to be toasted per hour, the size of toaster, and electrical data.

8

Cleaning and Sanitizing Equipment

Dishwashers

Dishwashing equipment should have the approval of the National Sanitation Foundation (NSF). Most manufacturers use 16-gauge stainless steel for the body and hood of dishwashing units; some use 18-gauge stainless steel for wash chambers. When purchasing a unit, specify model number, conveyor direction (left to right or right to left), and type of utility (electricity, gas, or steam). If electricity is selected, specify voltage and phase. Thermometers, which are standard on most models, should be located where they can be seen easily.

Other items that are usually standard but should be considered are revolving wash arms, rinse arms, and stainless steel strainer pans, which are easily removed for cleaning; a spring-counterbalanced door on single-tank, rack-type equipment; and an automatic timer for wash and rinse cycles. Most machines are interwired at the factory, so connecting them to a junction box is usually all that is necessary to install the electrical parts. However, this feature should be specified when purchasing the equipment.

PREWASH UNITS. There are four types of prewash units that can be added to the single-tank conveyor unit to improve the dishwashing procedure:

1. A 22- or 36-in. fresh water prewash unit with powerful

119

jets that strip food from tableware uses 110–140°F (45–60°C) fresh water, so the daily water cost is increased considerably.

2. A 22-in. recirculating prewash unit recirculates overflow water from the dishwasher instead of fresh hot water, so there is no additional water cost. This unit has a motor rated at ¼–½ hp.

3. A 36-in. recirculating prewash unit with a 1-hp motor does an excellent job of removing soil from dishes.

4. A prewash unit that can be combined with the double-tank unit or made a part of any flight-type unit.

SINGLE-TANK WASHERS. The smallest dishwashing unit useful for a food service facility is about 26 × 28 in. This unit includes wash and rinse cycles, and standard equipment usually includes the wash and rinse thermometers and an electric timing panel. This model can be obtained with a motor rated at ½ hp or 1 hp. Tank capacity ranges from 18 gal to 23 gal. Rinse water flows into the wash tank, so a constant flow of detergent is needed.

The door opening can range from 15 in. to 19 in. in height, so consider this factor in relation to the tray size used. A higher opening can be specified but costs extra. Some models have revolving wash and rinse arms above and stationary wash and rinse arms below; this combination cleans the dishes better than stationary arms above and below.

A single-tank unit with doors is adequate for a facility that serves up to 50 meals at one time. About 1 min is required for each rack of dishes to wash and rinse, but some presoaking is necessary with this type of unit. Because each rack remains in the unit about 1 min and because there would be about 30 racks of dishes, glasses, and silverware, it would take at least 30 min to wash the dishes for fifty people in a single-tank unit with doors (Fig. 8.1).

SIDE LOADERS. The side loader has been designed to save space in the dishwashing area. The dish racks are automatically indexed 90° from the soiled-dish table into the machine by means of a steel reciprocating, center-indexing panel. The racks must be equipped with both center and side index trips. When this type of unit is used because of limited space, a cart

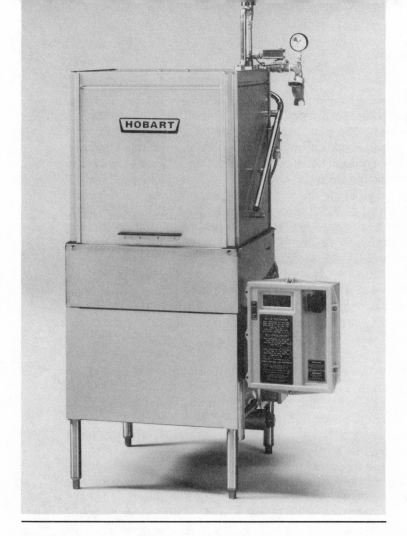

Fig. 8.1. Single-tank dishwashing machine. (*Courtesy Hobart Corp., Troy, Ohio*)

or table for scraping and racking dishes must be provided.

SINGLE-TANK CONVEYORS. The next size unit is the single-tank conveyor, which ranges from 36 in. to 54 in. in length. It has a 1- or 1¾-hp motor, and the conveyor mechanism has a ¼-hp motor. This unit has curtains at either end because the racks are automatically pushed out onto the clean-dish table at the end of the cycle, eliminating the need for opening doors and manually removing the racks. This unit includes a power wash and a final rinse. The height of the opening varies from 16¾ in. to 18 in., a factor to be considered if 18 × 26 in. bun pans are to be washed.

Splash shields mounted at each side of the front end are standard equipment on some units of this size. These shields prevent water from the racks and curtains from splashing on the floor. A single-tank unit with curtains usually has a conveyor speed of 3.5 ft/min and can easily handle dishes for a 125-bed health care facility or 150-seat restaurant. With two racks of dishes passing through the 44-in. unit at one time, approximately 75 racks of dishes, trays, silverware, and glasses could be washed in 35–40 min. The dishes are exposed to a power wash for approximately 1 min; a final rinse flows into a separate tank (Fig. 8.2). The addition of a prewash unit eliminates the need for prewashing dishes and therefore saves labor.

DOUBLE-TANK CONVEYORS. Double-tank units vary in length from 64 in. to 81 in. The motor for the wash tank usually is 1½ hp, and the motor for the conveyor is ¼ hp. Longer units have a 2-hp motor for the power wash and the power rinse units. There is better separation of the wash and rinse areas in the longer units, reducing water transfer between the wash and rinse compartments. The larger units include a power rinse as well as a final rinse for complete removal of detergent from dishes, plus a longer time for the 180°F (80°C) water to hit each dish. This provides better conditions for the dishes to air dry when they come out of the unit.

Consider a double-tank unit or a single-tank conveyor unit plus a prewash unit for a facility serving more than 150 meals per meal period. This unit provides more thorough washing of every dish and also speeds the operation. The addition of a prewash unit eliminates some prewashing by hand and results in cleaner dishes.

FLIGHT-TYPE UNITS. A flight-type unit costs about 30% more than a rack unit with a similar combination of tanks and motors, and it will wash about 35% more dishes in the same length of time.

The motors for the conveyor drive in rackless units are ¼–½ hp. Other motors vary according to the kind of tanks

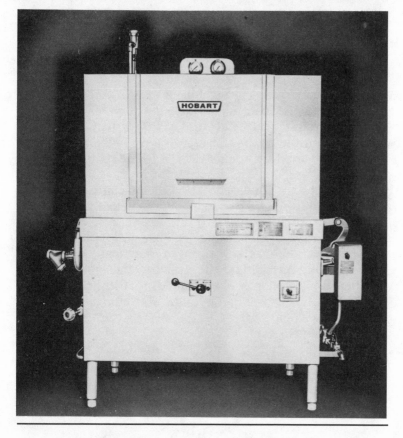

Fig. 8.2. Single-tank conveyor dishwashing machine. (*Courtesy Hobart Corp., Troy, Ohio*)

that are combined for this unit. Larger machines have larger tanks and require motors of about 3 hp.

Most units include a recirculating prewash, a power wash, a power rinse, and a final rinse unit. The loading section is usually about 36 in. long. Draining, drying, and unloading extensions are available in lengths of 4, 5, 6, 8, and 11 ft. Total length can range from 14 ft to 26 ft.

Some units have openings from 29⅓ in. to 30⅞ in. wide and about 24 in. high. This is excellent if the 18 × 26 in. bun pans and other large utensils are to be washed in the machine. A safety mechanism at the end of the unload section is necessary to stop the conveyor automatically if the dishes are not removed immediately.

Consider using a flight-type unit when numerous plates of various sizes, saucers, vegetable dishes, and trays are to be washed. The cups and glasses still have to be racked and the flatware washed in a rack or in cylinders placed in a rack.

AUTOMATIC TRAY WASHERS. The machine takes the sectioned tray (15/min) and while inverting it, does a thorough prewash scraping through a special "waterfall" system that flushes the soil from the tray with water of 130–140°F (55–60°C) removing all food waste and debris, including paper products.

The food scraps are conveyed out of the machine on a perforated self-cleaning endless belt and deposited in a waste container. If flatware is accidentally left on a tray, it is magnetically separated from waste and deposited in a flatware soak tank. The tray is then inverted and proceeds through the wash and rinse tanks. The clean tray then goes to the dryer where a high-velocity airstream strips all rinse water from the tray. The trays are automatically counted and loaded in three stacks on a cart.

ACCESSORIES. Some features have to be specified at a recognized extra cost.

1. A soak sink is necessary in any dishwashing layout. A portable soak sink provides flexibility and can be located to suit different employees' needs.
2. A flexible hose and a drain must be available for filling and draining a portable sink.
3. A disposer-trough combination is an excellent addition to the soiled-dish area if the labor load requires two or more persons. There is a water supply at one or both ends of the trough. The disposer may be located either at one end or in the center of this unit. The trough, which is usually 6–8 in. wide, slopes from the water supply to the disposer. A guard is usually provided to prevent flatware and other small items from getting into the disposer.
4. A slanted shelf that is 20 in. or multiples of 20 in. in length set at a 25° angle about 15–17 in. above the work surface provides easily accessible space for cup and glass racks. Provide shelves below the counter on the soiled-dish side for storage of dish racks. The space below the counter on the clean side should remain open for storage of mobile dish carts.
5. A rack return should be provided with a rack-type machine. Sections of rollers the width of the dish racks may

be used. The trough for these rollers should not drop more than ¼ in./ft because it is difficult to lift the racks out of a deep trough. Sections of rollers 5–6 in. wide could be mounted on adjustable feet and could be located on the counter at the back of the dishwasher. Be sure a guard rail is attached to prevent racks from turning over.

6. Extra dish racks are necessary if tableware is to be stored in the racks. Manufacturers furnish four racks with the machine. These heavy racks are difficult to handle, and plastic racks that weigh less and are easier to handle are available. Consider the particular needs of the facility when purchasing additional racks. There should be a minimum of three tray racks, three bowl racks, and three plate racks. The quantity of cup and glass racks depends on the number of cups and glasses to be washed and stored. Special racks with open ends are needed to wash trays that are longer than 20 in., which is the standard rack size.

7. Booster heaters are necessary to provide 180°F (80°C) water for the final rinse. The booster heater should include a strainer on the steam and hot water lines, an electric thermostat for actuating the solenoid steam valve, a hot water pressure regulator, and a pressure gauge. Close mounting of the booster heater eliminates line temperature loss and ensures a constant supply of 180°F (80°C), sanitizing final rinse water. There is a pencil device available that melts at 182°F (82°C), which may be used to check for proper rinse temperature in a dishwasher.

8. A door safety switch prevents the door of the dishwasher from opening while the machine is in operation.

9. A pressure gauge is used to determine the pressure of the water entering the manifold. The water pressure should be between 15 psi and 25 psi. If the water pressure is too low or too high, the dishes are not cleaned properly.

10. A water pressure–reducing valve reduces water pressure that is too high.

11. A steam pressure–reducing valve reduces steam pressure that is too high.

12. A low-water protection device prevents heaters from burning out when water is low in the tank.

13. A rinse economizer eliminates use of 180°F (80°C) water when no dishes are passing through the rinse area in the flight-type equipment.
14. A stainless steel front enclosure panel encloses the motor, pumps, and frame and reduces the need for cleaning an area not easily cleaned.
15. Removable stainless steel inspection doors make it possible to clean the scrap trays and wash and rinse the arms easily. This type of door usually locks in a raised position.
16. Interplumbed drains and a motor control unit are required on multiple-tank units.
17. Stainless steel legs are almost a necessity in an area that is moist for extended periods of time. Other materials would rust and need to be replaced.

MOLDED DOLLIES AND BUSING CADDIES. Molded dollies are designed so that cup or glass racks can be stacked on them. They are usually dentproof, rustproof, and sturdy enough to hold three full cup racks or two full glass racks (more empty racks can be held). The unit in Figure 8.3 is 20½ × 20½ × 5 in. and will hold up to 1½ gal of liquid. Dirty cups or glasses from dining tables can be placed directly into a dishwasher rack and any liquid will drain into the dolly below the rack. Dollies can be mounted on 4-in. casters and have a handle attached, or it can be set on a utility cart and rolled to the dishwashing area. Some models are available without casters.

Fig. 8.3. Molded dolly on casters. (*Courtesy Raburn Products, Inc., Wheeling, Ill.*)

Fig. 8.4. Busing caddy. (*Courtesy Caddy Corp. of America, Pitman, N.J.*)

Busing caddies (Fig. 8.4) allow for efficient cleaning of the dining area when the meal is finished. The cart can have pans for soaking plates, a pan for waste food, a trash can on one end, molded dollies under the cup and glass racks, cylinders in pails for sorting and soaking flatware, and space for empty trays. Busing carts can reduce the time for clearing a dining area by 50%; they can, in turn, be rolled directly into the dishwashing area for unloading into the washers.

VENTILATION FOR DISHWASHERS. Built-in ventilating hoods usually increase the efficiency of the dishwashing operation, because the reduction of humidity results in more comfortable working conditions and increases the speed of air drying.

Stainless steel ventilating cowls can be fastened at the

wash and rinse ends of a conveyor dishwasher to exhaust steam. These cowls add about 16 in. to the overall length of a unit. A vent in each cowl connects to an exhaust system. Dampers installed in the duct system will prevent too rapid exhausting, which could chill the final rinse water.

A power-driven blower dryer may be a good addition if there are numerous plastic trays and/or plastic dishes to be dried. These units, which have a full hood, are approximately 5 ft long.

Warewashing Facilities

Sanitizing pots and pans can be done effectively by hand in a properly designed 3-compartment sink with drain boards at each end in all but very large facilities. A good working height is 36 in. from floor to front rim. Each sink should be 20–24 in. from front to back and 24–30 in. wide. Most employees can wash pots without undue stooping if the wash sink is not more than 14 in. deep, but a 12-in.-deep sink is better. A sink that is 36 in. high and 14 in. deep makes the bottom of the sink, or the working surface, 22 in. from the floor.

Soak and rinse sinks may need to be deeper than the wash sink. When changes are made from the standard size, the special fabrication increases the expense. Consider the use of a portable soak sink to cut expenses. A floor drain is necessary if a portable sink is used.

Faucets have to be specified and purchased as a separate item from sinks and basins. There are many different kinds that can be used on a pot-and-pan sink. To mention a few, there are the

Deck mixing faucet with 6-in. cast nozzle
Service sink nozzle with top brace, vacuum breaker, and
    ¾-in. hose thread
Workboard faucet with swing nozzle
Pot- and kettle-filling nozzle

It is important to specify a nozzle that is long enough to swing over the sinks that it is supposed to fill and to have a hose thread if a hose is to be used from that water source.

Include a disposal unit of at least 1 hp in the area where soiled pots and pans are scraped, sorted, and stacked or racked. Provide a cone so that the waste goes into the disposal unit rather than spreading over the flat counter.

The last sink compartment is used for the final rinse and needs extremely hot water. This is generally supplied by an immersion heater below a false bottom. This heater is a permanent installation, so work flow must be determined before it is installed.

The warewashing sink should be made of 14-gauge, nickel-bearing stainless steel. Drainboards should be pitched to drain well and all corners should be coved, welded, and seamless to comply with the NSF code. You have to specify faucets, which cost extra. Lever waste handles also should be specified to eliminate handling stoppers, especially in the rinse unit where the water is very hot. You need a drainboard at least 4 ft long for clean items so that the pots and pans can air dry. One or more mobile pot racks are necessary so that they can be moved to the cooking area and back to the wash area as needed. These pot racks need shelving with smooth slats or openings to allow air circulation and easy cleaning.

A unit mounted in the pot sink for circulating the wash water assists in removing heavy soil and makes the job of warewashing much easier. A pump inlet and outlet is cut into the side of the sink, and the constant motion of the water helps to remove soil and keep it in suspension. Specify left- or right-hand mounting so that the grease line and switch are installed in the correct side. The unit can be installed on any stainless steel or galvanized sink that is at least 14 in. deep. The motor plugs into a 120-volt outlet.

In large facilities, a pot-and-pan washing machine may be necessary. Counter height units use a rack that is filled with pans and pushed into the washing chamber. The rinse cycle takes place in the same chamber. Floor units are used to sanitize racks, serving carts, and any other items that can be wheeled into the chamber. The unit must be countersunk so that the floor of the machine is flush with the floor of the room.

Food Waste Disposal Units

A food waste disposal unit will increase efficiency and economy and will reduce odor and bacterial contamination regardless of the size of the operation. Units may be designed as trough, sink, or cone installations.

When a disposal unit is purchased, choose its placement carefully. Disposal units frequently are installed in sinks because this is the common usage in households. However, placing a disposal unit in a sink in large-volume operations removes the sink from use for other purposes during much of the day. A better placement is in a table or drainboard. A collar that rises at least 1 in. above the counter top prevents food from splattering while the machine is in operation. The disposal must be designed so that when it is not in use the collar can be removed to make a flat work surface.

First identify the locations where disposal units are needed. The areas that produce the greatest volume of waste are dishwashing, vegetable preparation, and pot-and-pan washing. Depending on the size of the facility, the kitchen layout (distance of one work unit from another), and the type of food service operation (conventional vs. convenience), disposal units also may be needed at the salad preparation area and near the cooking units. If the kitchen has only one disposal unit, the best location is in the soiled-dish area. It may be more economical as well as more sanitary to place disposal units at all locations where waste originates than to carry the waste from one location to another in the kitchen. At the location of each disposal in the kitchen, post operating instructions, safety precautions, and a list of items that should not be put into the disposal. Adequate instruction for all kitchen personnel will reduce breakdown and maintenance problems.

Disposal units of ¼–½ hp developed for household use are not satisfactory for institutional use. A restaurant that serves 100 people during a meal period or a 50-bed health care facility that uses one disposal unit to grind all kitchen waste should choose a disposer that has at least a 1-hp motor. A 200-bed health care facility or a restaurant that serves 400 people during a meal period probably will require three disposal units: a 2-hp unit at the soiled-dish table, a 1-hp unit at the vegetable and salad preparation area, and a ½-hp unit at the pot-washing sink. If prepared fruits and vegetables are used, a disposal unit may not be needed in the vegetable and salad preparation area.

It is a mistake to buy a disposal by horsepower alone; the horsepower rating may be too high in relation to other factors. Think carefully about the rotor diameter and the size of the body, also. Additional horsepower only enables the disposal to handle heavier wastes; it does not increase the output. Larger rotors and bodies promote more centrifugal force to grind waste and to speed removal. For instance, a small, 1-hp motor used in a disposal for a soda fountain or snack bar may need only a 6½-in. rotor, but this unit would not be suitable for anything except light waste loads. In large-volume operations, an 8-in. diameter rotor with a 2-hp motor, or a 10-in. diameter rotor with a 3- to 5-hp motor, or an 11-in. diameter rotor with a 7½-hp motor is recommended.

An operating disposal is noisy, so to muffle the sound as much as possible, disposals can be equipped with rubber-cushioned mountings and insulation. A disposal supported on legs creates less noise and vibration than one attached to the underside of a sink or table. If the disposal is supported on legs, there must be clearance for mopping the floor underneath the motor.

When purchasing a new disposal be sure to specify resilient mountings to eliminate vibration above and below the grinding chamber and to reduce noise. A completely enclosed motor is necessary to keep out moisture. A prebreaker may be added to reduce fruit rinds and similar wastes to smaller sizes before grinding.

Ask about the guarantee, and find out how replacement parts can be obtained. Check on whether service is provided locally, by the suppliers, or by the manufacturer and how much time must be allowed for repair service.

Handwash Basins and/or Lavatories

It is mandatory to have a handwash basin and/or a lavatory in every food service department, and it is important to have at least one handwash basin in every area where both dirty and clean items are handled, such as the dishwash area. It is also important to provide a means of turning the water on and off without touching the handles of the faucets: wrist-action handles, knee-action handles, or foot pedals; short handles that have to be turned on and off by hand provide a chance for recontamination.

Disposables

The degree to which disposable tableware is used depends on the philosophy of the facility. Some health care facilities use disposables for nourishments or snacks because they are lightweight and easy to handle. Schools that operate from a centralized kitchen and transport bulk or portioned food use disposables to eliminate the need for a dishwasher in the satellite kitchen.

If a completely disposable system is chosen, no dishwashing equipment is needed. However, there are many times when some glasses, flatware, serving utensils, knives, spatulas, and other small equipment are used. A means for washing and sanitizing these items is then necessary and a dishwasher may be the best answer.

The use of disposables may eliminate some labor, but unless the number of employees can be reduced, a false economy will be created. Labor is needed to transport disposables to the serving area and remove used products. Shredding, compacting, or incinerating equipment is needed to dispose of soiled tableware.

A great deal of storage space is needed for disposables. If storage is at a premium it may be necessary to have additional deliveries, which increases the cost of handling and billing. There is also the cost of waste in disposables when clean items are spilled upon or allowed to be exposed to dust and dirt.

Each food service will need to make a comprehensive study to determine whether disposables or permanent ware best suits its needs and budget.

9

Small Equipment

Pans

Pans do not come as a part of the hot food serving units, refrigerators, ovens, or other equipment; therefore they must be ordered separately. When ordering food storage or serving pans, or pans for any purpose, the use will determine several factors that should be considered:

1. Gauge indicates the weight of the metal; the heavier the gauge the less likely it is the pans will warp, especially pans used for oven cooking. These pans are usually available in 18, 20, and 22 gauge, 18 being the heaviest and 22 the lightest. The gauge is usually stamped on the bottom of the pan.
2. Size varies according to need, with 12 × 20 in. pans fitting the standard opening. However, pans are available in one-half size (12 × 10 in.), one-third size (12 × 6⅔ in.), one-quarter size (6 × 10 in.), one-sixth size (6 × 6⅔ in.), and one-ninth size (4 × 6⅔ in.). Smaller pans provide enough space for many special items. There are 6 × 20 in. pans that also fit the standard opening.
3. Depth of a pan may be 2½, 4, or 6 in. The 2½-in.-deep pan is excellent for serving vegetables such as broccoli, asparagus, and cauliflower, as well as casseroles such as macaroni and cheese, meat loaf, or tuna noodle bake—all of which look nicer when there is a crisp top for each serving. Deeper pans are better for potatoes, gravy,

soups, and stews. The capacity of a pan and the number of portions it will serve is stamped on the bottom of most pans.

4. Nesting or stacking ability is an important feature in storage of clean pans. The bottom of the pan is a little smaller than the top; therefore the pans can be stacked easily without wedging together.

5. Perforated pans are made of stainless steel and are designed to fit into standard full- and half-size pans. The perforations allow moisture or grease to drain off the food into the solid pan underneath.

6. Covers may be flat, flat hinged, or semidome hinged. Most of the covers are 18-gauge stainless steel and have some kind of handle for lifting them. Usually only one cover is needed for each well in the hot food serving table.

7. Pan adaptor bars are made of 18-gauge stainless steel and fit across the 12-in. width of the opening of the hot food well to support the various combinations of small pans.

Bun pans are usually 16-gauge aluminum and are 18 × 26 in. and 1 or 2 in. in depth. Many are needed for storing portioned salads and desserts in the refrigerator, for holding portioned room-temperature items such as cake and cookies, for use in the bakery, and as carriers to be placed on top of utility carts.

Double boilers may be either stainless steel or aluminum and range in size from 2 qt to 25 qt, depending on the type and manufacturer. It is convenient to have several sizes, especially if there is no steam-jacketed kettle available.

Mixing bowls are more durable if constructed of stainless steel. They are available in sizes ranging from ¾ qt to 30 qt. The sizes purchased will depend on the size of the facility.

Saucepans are constructed of heavy-duty aluminum. They are available in sizes ranging from 1½ qt to 10 qt. Some lids are necessary but much stove-top cooking is done in open pans.

Frying pans are either cast iron or aluminum clad in silverstone to form a nonstick surface. They range in size from 6½ in. to 15¼ in., with or without lids.

Roasting pans, layer cake pans, pie pans, angel food cake pans, and muffin pans may be needed, depending on the kind and size of the facility.

Dish Racks

Standard racks are 19¾ in. square. There should be a four-way index so that they will ride through the dishwasher regardless of the side that is pushed in first. These racks should stack easily. Plastic racks are much lighter and easier to handle than plastic-coated wire racks, and there are special racks for many types of dishes and trays.

Glass racks need to be taller than cup racks; the overall height is usually 5½ in. Very tall glasses need to be washed in racks with specially designed individual compartments. To reduce handling of each glass enough racks are needed for all the glasses in use.

Cup racks can be flat or divided. One important feature is a plastic ledge about 1 in. high located diagonally across each compartment of the cup rack to tilt the cup so that all the water will drain off the bottom. There should be enough cup racks to store all the cups in use daily.

Tray racks usually hold eight trays. Some trays are longer than the standard rack, so an open-ended rack with strong dividers has been designed for washing trays in a dishwasher. A minimum of two of these racks is needed so that trays can drain in one rack while another is loaded.

Half-size racks are available for use where storage and handling is a problem.

Knives

The quality of a knife depends on the blade, the handle, and how well the two are joined. The basic ingredients of all knife blades are steel and carbon. When chrome is added, the result is a stainless steel blade. Sometimes vanadium alloy is added to carbon steel to make the blade stronger and help it hold its cutting edge. The quality of the blade depends not only on what it is made of but also on its heat treatment and the tempering process during production.

A carbon steel blade has the finest cutting edge and is usually the choice of professional chefs, but it will be affected by rust and food stains and will have to be sharpened oftener. Stainless steel can produce a sharp and long-lasting cutting edge, but not as sharp as carbon steel. A good blade will bend and snap back into place immediately. The cutting edge can have a variety of forms: straight, serrated, scalloped, or wavy. Slicing knives can have scalloped or straight edges.

Knife handles are made of maple, rosewood, walnut, plastic, plastic-wood combinations, bone, and ivory. Plastic-impregnated wood makes a very durable, dishwasherproof handle.

The tang (that portion of the blade that is fitted into the handle) should extend the full length of the handle and should be held in place by three rivets. If the tang is too short or is not well fastened, it will loosen with use.

A French knife is excellent for chopping and mincing. It tapers evenly to a sharp point from a wide base where it is joined to the handle. These knives are available in lengths from 6 in. to 12 in.

A paring knife is used to pare fruits and vegetables, remove eyes from potatoes and blemishes from fruits, or pit fruits such as peaches and plums. The blade is about 3 in. long. It is available in various widths, curves, and end shapes.

A utility or trimming knife is used to slice fruits and vegetables, core fruits, and bone and trim meat. The blade is 5 or 6 in. long, and it may be serrated, scalloped, or straight.

A carving knife is used to carve hot meats. The blade is long and fairly wide and ends in a curved point. The edge can be scalloped or straight. It is usually 8–12 in. long.

A vegetable peeler has two blades and has knee action, which means that the blades operate as a pair and rotate on a common axis. It is used for limited amounts of paring.

A knife rack is the key to protecting the knife blades.

Other Utensils

A variety of utensils, in addition to knives, are used in the preparation of food in food service facilities.

Spoons are usually made of 18-gauge stainless steel. They are available with slotted, solid, or perforated bowls, which come in 11-, 13-, or 15-in. lengths. The handles may be stainless steel or may be clad in a heat-resistant composition material.

Ladles are generally made of 18-gauge stainless steel and are available in 1-, 2-, 4-, 6-, 8-, or 12-oz sizes.

Spoodles are available in the same sizes as ladles, with perforated or solid bowls and sanitary handles that are heat resistant and shaped so that it is easier to serve (Fig. 9.1).

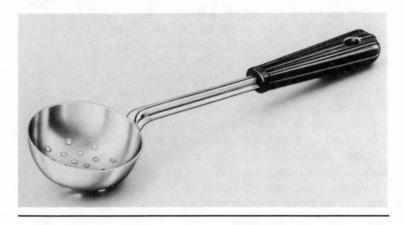

Fig. 9.1. Portion-control spoodle. (*Courtesy Vollrath Co., Sheboygan, Wis.*)

Wire whips are efficient for stirring foods such as gravies and sauces that have a tendency to lump. Each wire is a cutting edge, so the whip offers many times the action of a spoon. They are available in three weights: fine piano-wire whips are used for light sauces and other light stirring; French whips are average weight and used for the most products; and heavy wire whips are used for stirring heavy batters and sauces. The most desirable length will depend on the amount of food to be stirred and the size of container.

Metal spatulas are available in a variety of lengths and widths and can be straight or bent. Use a long, flexible, straight spatula when frosting cakes. Use a short spatula with a blade 1½ in. wide and 4 in. long to spread butter, mayonnaise, or soft fillings on sandwiches. Use a spatula with a bend just below the handle for serving sheet cakes and gelatin salads. Select one that is the same width as the piece of food being served. Special pointed spatulas are available for serving pie and layer cakes.

Rubber scrapers are the most efficient way of avoiding loss of a food substance when transferring it from one container to another. They may be crescent-shaped for cleaning out steam-jacketed kettles; large for scraping large containers; small for scraping small containers; and narrow for cleaning out narrow-mouthed bottles.

Tongs are available in several lengths and styles. The most popular style for food handling is pom tongs. They may be 6, 9, or 12 in. long. They are useful in serving preportioned meat, ear corn, spinach, relishes, and ice.

Scissors work more efficiently than knives in some instances, particularly for cutting dates, marshmallows and other sticky foods, and meat and vegetables for salads.

Egg slicers may be used for slicing eggs neatly and quickly. They are also handy for cutting cooked carrots and potatoes, butter, margarine, bananas, and other soft foods.

Colanders, china caps, and strainers are used for draining and straining. They are available in a wide variety of sizes, depending on need.

Measuring spoons, can openers, bottle openers, cooking forks, ice picks, skimmers, biscuit cutters, cookie cutters, and rolling pins may be selected, depending on the kind of facility and the menu.

Purchase heavy duty, well-constructed tools. Keep them sanitary and in good repair. Select the correct tool for each job that is to be done. Plan a definite place for tools and be sure to return them to that place after each use.

10

Making the System Function

Checklist for Building or Remodeling

To make sure that all necessary space relationships and equipment have been considered, check the following items.

1. Food service supervisor's office.
   a. Possible to observe surrounding areas from the office.
   b. Good ventilation and lighting.
   c. Equipped with a desk and a telephone and maybe an intercommunication system.
2. Receiving area.
   a. Dock protected from weather by an overhang.
   b. Located near the storage area.
   c. Equipped with oversize delivery doors.
   d. Scales.
   e. Carts and/or trucks for moving supplies.
3. Storage area.
   a. Adequate shelves for bulk storage.
   b. Shelves adjustable for the height of items.
   c. Space for disposables near the area of use.
   d. Security provisions (locks).
   e. Adequate ventilation and lighting.
   f. Accessible to the production area.
   g. Cart with a manual or an electric can opener and a storage shelf below.

4. Refrigeration facilities.
   a. Adequate refrigerator and freezer space to accommodate the production and serving systems used.
   b. Security provisions (locks).
   c. Accessible to the receiving, production, and serving areas.
   d. Food files for trays of prepared salads and desserts and extra pans and racks, as needed.
   e. External thermometer for each refrigerator and freezer.
5. Food preparation area.
   a. Adequate 3½- to 4-ft aisles.
   b. No oven and refrigerator doors opening opposite each other.
   c. Adequate equipment for type of menu.
   d. Minimum cross traffic.
   e. Cook's table with drawers, a knife rack, and a cabinet above; baker's table with drawers and mobile bins with a cabinet above.
   f. Mobile tables with food file shelving for the salad area in addition to built-in counter space with sink.
   g. Located close to distribution points for prepared food.
   h. Fire extinguishers and an adequate ventilation system.
6. Food distribution area.
   a. Space for tray setup including an assembly unit for trays, tray covers, napkins, flatware, condiments, and perhaps bread plates and saucers.
   b. Provision for keeping food hot during service.
   c. Storage for dishes with equipment for heating plates and perhaps beverage pots and a mobile unit for trays, etc.
   d. At least 5-ft aisles for large tray carts, portable hot tables, etc.
   e. Convenient access to corridors, elevators, or dining room.
   f. Five or six electrical outlets of proper voltage for food service equipment.
7. Warewashing area.
   a. Located close to the production area.
   b. Large disposer with cone in counter area preceding

soak sink, wash sink not more than 10–12 in. deep, rinse or sanitizing sink 14–16 in. deep with heating element, levers for drain release below sink and adequate space (about 4 ft) on clean end for air drying of pots and pans.

   c. Adequate racks for storage of clean utensils.

8. Dishwashing area.
   a. Adequate ventilation and lighting.
   b. About 2 ft² per person served. (This figure will fluctuate after certain minimum needs are met.)
   c. Sufficient garbage cans or grinders and trash receptacles.
   d. Portable soak sink, shelf above dirty area for two dish racks for cups and glasses, space between disposer and dishwasher for two dish racks (about 42 in.).
   e. Rack return if possible (if rack-type machine).
   f. Between 6-ft and 12-ft clean-dish counter so that there is space for several racks of dishes to air dry at one time.
   g. Provision for handwashing facilities.
   h. Located close to dining area.
   i. Floor drain in dishwashing area.
   j. Plenty of width next to clean-dish counter so that there is room for at least one piece of each kind of mobile dish storage equipment to be used.
   k. Flow pattern in dishwashing area so that carts continue through the area rather than having to turn or back out.

9. Employees' facilities.
   a. Handwashing facilities.
   b. Adequate dressing rooms and lockers.

10. Equipment needed but not usually included with the major items unless specified.
    a. Faucets for sinks.
    b. Pot fillers for kettles near ranges.
    c. Remote-controlled compressor for refrigerators if not self-contained.
    d. Thermometers with refrigerators.
    e. Additional racks, roll-out shelves, food file shelving, locks for refrigerators as specified.
    f. Fans for hoods, good ventilating system.

g. Cords for toasters.

h. Backs or fronts on parts of equipment that are exposed, such as the motor of the dishwasher.

i. Plywood or other insulation materials under stainless steel tops for deadening of noise.

j. Pans for hot food serving counters.

k. Steam-reducing valves.

l. Booster heaters for dishwasher and rinse sink of pot-and-pan wash unit.

m. Additional racks for dishwashers.

n. Solenoid valves with food grinders.

o. Extra cylinder containers for flatware.

p. Racks for attachments for mixers.

q. Locks for two of the four casters on most mobile equipment.

11. Clock and bulletin board needed in every kitchen.

12. Electrical outlets.

a. Toasters, hot food carts, heated dish dispensers, in some instances cold food serving equipment in serving area.

b. Mixer in baking and cooking area.

c. Hot carts in storage area, on serving line.

Checklist for a New Food Service Department

To provide a smooth procedure for the initial operation of the new facility the following steps should be considered.

1. See that all large equipment is hooked up properly and working well.

2. See that the following items are on hand:

a. Racks, pans, and storage baskets for refrigerators.

b. Beaters, whips, and bowls for mixers.

c. Inserts for hot food serving tables.

d. Pans for portioned salads and desserts.

e. Carts for use in tray service.

f. Dish racks for cups and glasses, racks for washing trays (open-end racks if trays are more than 18 in. long), and cylinders for washing flatware.

3. Demonstrate use of equipment to employees and have employees try out the equipment.
4. Write job descriptions that are suited to the new facility.
5. Explain every detail of the new job descriptions to employees.
6. Write rather simple menus to be used for the first week of operation.
7. Purchase food for the first week of operation.
8. Plan and serve a meal for some employees or other group to test the equipment and coordinate the work activities.
9. Set up a file (in the food service department) of operators' instructions for all new equipment.
10. Set up a procedure for the use and care of all equipment.

Preventive Maintenance

A preventive maintenance program saves money and time. Maintenance and cleaning are integrated functions. Equipment kept clean and oiled will be less apt to develop serious problems than that not cared for properly. When machines are cleaned and inspected regularly, small problems will be discovered and necessary repairs made before costly, time-consuming breakdowns occur.

To be effective, maintenance inspection must occur on a regular basis as does the cleaning. The system for keeping records about the equipment, installation and operating instructions, guarantees, and repairs can be as simple or as complicated as you like. A card can be used for most of the information you will need (Fig. 10.1), or you may prefer to have similar information on a full sheet of paper so that it may be filed with the installation and operating instructions, or you may choose to develop a notebook that contains this information.

EQUIPMENT RECORD

Item of equipment_____

Manufacturer_____Model No._____

Address_____Motor No._____

Purchased from_____Serial No._____

Purchase date_____Guaranteed for,_____ Cost_____ New___
                                                          Used___

Capacity_____Attachments_____

Date_____Description of repairs_____Cost_____

_____

_____

Fig. 10.1. Equipment maintenance record form.

BIBLIOGRAPHY

Federal Energy Administration. *Guide to Energy Conservation for Food Service,* No. 041-018-00085-2. Washington, D.C.: Government Printing Office, October 1975.

Hitchcock, Mary L. *Food Service Systems Administration.* New York: Macmillan, 1980.

Hoffman, C. J., and Zabik, M. E. "Current and Future Food Service Application of Microwave Cooking/Reheating." *J. Am. Diet. Assoc.* 85(1985):929.

Hysen, P. "Guide to Successful Planning." *J. Am. Hosp. Assoc.* 48(March 1974):126–32.

Jernigan, A. K. "Selection of Mobile Dish Storage Equipment." *Hospitals* 1970(January 16):61–62.

_____. "The Care and Feeding of Waste Disposals." *Hospitals* 1971(August 16):86–87.

_____. "Selection of Dishwashing Equipment." *Hospitals* 1971(November 1):91–93.

Jopke, Walter H., and Hass, Duane R. "Contamination of Dishwashing Facilities." *Hospitals* 1970(March 16):124–27.

Kazarian, Edward A. *Food Service Facilities Planning,* 2d ed. Westport, Conn.: AVI Publishing Co., 1983.

Kotschevar, Lendal H., and Terrell, Margaret E. *Food Service Planning, Layout and Equipment.* New York: John Wiley & Sons, 1985.

McCarthy, Robert. "FM Design—Bloomington Hospital, Kitchen & Employee/Visitor Cafeteria." *Food Manage.* 1985 (April):158–59.

McCool, Audrey Carol, and Posner, Barbara Millen. *Nutrition Services for Older Americans: Food Service Systems and Technologies.* Chicago: Am. Diet. Assoc., 1982.

145

Midwest Research Institute. *Energy Management and Energy Conservation Practices for the Foodservice Industry.* Chicago, Ill., February 1977.

National Fire Protection Association International. *Ventilation of Cooking Equipment 1970,* Boston, Mass.

National Sanitation Foundation. *Standard No. 33 for Commercial Cooking Equipment Exhaust Systems,* Ann Arbor, Mich., 1970.

Ross, Lynne. *Work Simplification in Food Service.* Ames: Iowa State Univ. Press, 1972.

_____. *Metric Measurement in Food Preparation and Service.* Ames: Iowa State Univ. Press, 1978.

U.S. Department of Agriculture. *Guide for Equipping On-Site School Kitchens.* Food and Nutrition Service. Program Aid No. 1091. Washington, D.C., 1977.

Weintraub, Bernard S., and Kern, Harvey D. "Wet Grinding of Disposables." *Hospitals* 1970(November 16):70.

INDEX

147

158786

642.57
Jernigan   J55fo
Food service equipment.